MW00359076

A Diderot Pictorial Encyclopedia of TRADES AND INDUSTRY

Manufacturing and the Technical Arts in Plates Selected from "L'Encyclopédie, ou Dictionnaire Raisonné des Sciences, des Arts et des Métiers" of

Denis Diderot

Edited with an Introduction and Notes by
Charles C. Gillispie

In Two Volumes
Volume Two

DOVER PUBLICATIONS, INC.
New York

This Dover edition, first published in 1993, is an unabridged and unaltered softcover republication of the work originally published by Dover Publications, Inc., in 1959. The plates were selected and the introduction and notes were written by Charles Coulston Gillispie, and the index was compiled by the editorial staff of Dover Publications. Most of the plates have been reproduced facsimile size.

DOVER *Pictorial Archive* SERIES

Manufactured in the United States of America
Dover Publications, Inc., 31 East 2nd Street, Mineola, N.Y. 11501

Library of Congress Cataloging-in-Publication Data

Encyclopédie. English. Selections
A Diderot pictorial encyclopedia of trades and industry : manufacturing and the technical arts in plates selected from "L'Encyclopédie, ou dictionnaire raisonné des sciences, des arts et des métiers" of Denis Diderot / edited with an introduction and notes by Charles C. Gillispie.
p. cm. — (Dover pictorial archive series)
Reprint.
Includes bibliographical references and index.
ISBN 0-486-27428-4 (pbk. : v. 1). — ISBN 0-486-27429-2 (pbk. : v. 2)
1. Industrial arts—Encyclopedias. 2. Industrial arts—History—Pictorial works. I. Diderot, Denis, 1713-1784. II. Gillispie, Charles Coulston. III. Title. IV. Series.
T9.E472513 1993
670—dc20 92-31820
CIP

Table of Contents

Plates

Glass

Glass

The art of glassmaking goes back to prehistoric times. Its origin, like that of metallurgy or agriculture, is lost in antiquity. It is not even certain when the technique of blowing bubbles of glass through a pipe was first applied to making hollow vessels. But whoever has had to construct simple chemical apparatus in a school or college laboratory will have some feeling for the nice dexterity required of the glass blower. Glass is not typically a solid. It is rather a rigid liquid. Instead of suddenly melting like ice or iron at some definite temperature, glass softens throughout the mass when it is heated and stiffens on cooling, like butter or taffy. The blower must work his material rapidly since it is malleable for only a certain time as it cools.

During the late Middle Ages and the Renaissance, Europe looked to Italy for fine crystal and elegant mirrors, and particularly to the great Republic of Venice and the tiny duchy of Altara, near Genoa. So jealous of their art were the Venetians that in the 16th century any glassblower who took his skills abroad was liable to the death penalty, and the glassworks were shielded from the spying eyes of competitors by their isolated location on the island of Murano, where they were moved by a decree of 1291. The Duke of Altara followed a different policy. Wiser, or less powerful, he contented himself with royalties and exacted a lifelong percentage of the earnings of those of his glassblower subjects whom he permitted to take service in foreign lands.

The spread of Italian technique could not be prevented. It took root very widely in France, where the native glass industry had itself been held in high esteem since the Middle Ages. So much was this so that skilled glassworkers were in an extraordinary position. Under feudal society, every other industrial or commercial occupation was degrading. Not so the glass industry, in which top artisans held noble rank. They were not workers—*ouvriers*—and therefore commoners, but *gentilhommes-verriers* — gentlemen-glassworkers. Several explanations have been advanced for this privilege. It is sometimes said that the crown promised to respect the status of the Venetian and Altarese glass blowers when it induced the first of them to come to France. Unfortunately for this romantic story, however, the nobility of artisans in glass goes back to the fourteenth century, before the arrival of the Italians. It may have been that certain enterprising noblemen went into glassblowing and managed to carry their status with them rather than that glass blowers were ennobled out of appreciation for their art.[1]

By the 18th century the French glass industry centered in the Northwest in Normandy and Picardy, and in the Northeast in Lorraine. There were in addition small establishments supplying local markets scattered throughout the country wherever there were woods enough to supply fuel. Methods and organization were somewhat different in the two main regions, but both were engaged in the two chief types of glass production, hollow-ware for utensils and flat glass for window panes and mirrors. The plates that follow will illustrate the construction of a glassworks in Lorraine devoted to small items—a petite verrerie—and the production in it of a single goblet. Heavy glass—gross verrerie—will be illustrated from a coal-burning bottle factory near Paris.

Several methods were in use for the manufacture of plate for mirrors or window panes and of the ordinary grades of flat glass. Normandy produced "crown glass," made by pricking a bubble and spinning it open into a cylindrical disc. In Lorraine, on the other hand, the blower blew a cylinder which was slit and rolled open into a rectangular sheet of "broad glass." Both were methods for flattening bubbles into panes, an awkward problem at best. Eventually both were to be displaced by pouring or casting plates of glass. This was a French invention of the late 17th century. It was the first step on the way to liberating the industry from dependence on the skill and capacity of the individual glass blower and to replacing craftsmanship by machinery.

1 See an excellent recent history of the French glass industry, *La verrerie en France de l'époque gallo-romaine à nos jours* by James Barrelet (Paris: Larousse, 1954), p. 48.

Plate 209 A Glass Factory

The introductory engraving gives a general view of the main workshop of a manufacturer of hollow glassware in Lorraine. The plates that follow will develop each operation in detail.

In the center stands the glass furnace, smoke and flame escaping through the open vents. The glass itself is prepared inside in great pots. Gatherers (m) reach their blowing pipes through the openings to pick up "gathers" of molten glass from the pots. A master-blower (p) reheats a bubble of glass called, because it is at an intermediate stage, a "parison." To the right an assistant blower (n) shapes his parison by blowing and rolling it on a marble slab, the "marver." In this shop there is a division of labor. The gatherer passes the pipes to the assistant who does rough blowing and passes them on to the blower.

Work benches (o) are designed for the artisan to roll his pipe along the arms and preserve the round form blown in the soft glass as he trims and shapes it with pincers and shears. Blowing pipes (x) are cooling in a tub of water. The men toss scrap glass or "cullet" into receptacles (y) from which it will be reclaimed to be melted down in a later batch. Cullet was an essential ingredient, and at the rear (q) an apprentice cleans remnants of glass from a pipe into a big bin.

Along the top of the furnace runs an annealing oven or arch (aa), to which the small doors (c, d) high up on the side give access. In it finished ware is tempered by heat-soaking. The partition along the left background separates the shop from the sarosel *room, a warehouse into which the annealing arch opens (see Plate 225) at its cooler end. There the glasses are taken out and packed for distribution.*

Bending toward this partition at the left, is the fireman (t), who has charge of the fuel. The wood must be split fairly fine to maintain an even heat. A supply of faggots is stored across the rafters to be thoroughly dried by the heat of the workshop itself.

a a
a a
a a
r
a
a
e
e
e
e
b
e
c
d
d
m
f
f
h
h
h
t
g
i
s
o
z
p
&
y
x

A Glass Factory

Vol. X, Verrerie en Bois, Pl. I.

Plate 210 The Glass Furnace I

The purpose of each successive step in glassmaking will be clear if the general design of the furnace is kept in mind.

This is a bee-hive furnace. Its size will be apparent from the scale at the bottom. The old French pied *is approximately equivalent to the English foot, and the* toise *to two yards. Fig. 1 is a transverse cut-away of the furnace. Three pots (c) are set in recesses (b) and rest in place on the "siege" or platform. A fourth (d) is shown in cross-section. The blower makes his gather through the opening (e). The fire is built on a grate (o), and the heat pours upward to the oven through a flue (f). From the main oven a second flue (h) conducts heat to the annealing oven (i), built on top of the bee-hive. Small doors (m) give access for the finished utensils.*

Fig. 2 is a top view of the furnace cut away at the level of the annealing arch with its flue (b), air-vents (d) and doors (c—shown as m in Fig. 1). Finished utensils are put through these doors onto iron trays (f), which are then slid slowly along the tunnel (a—shown as y in Fig. 1) to the warehouse (not shown at the right). The heat decreases gradually along this tunnel, until at the end the ware has been cooled to normal temperatures. The rest of this diagram can be better understood by reference to Fig. 4.

Fig. 4 shows a cutaway of the furnace at the level of the main oven. At the center is the flue (c). Around the circumference are openings (m and n) through which the blowers work, protected behind small ramparts (t). Seven pots rest upon the hearth. Each contains a different composition. Green glass—the cheapest—will be made in one; brown glass for bottles in a second; clear glass for goblets in a third; glass for other products in the others. Passages (s) give access to the fire below from both ends.

Fig. 3 is an end-view of one of these passages. The fire is fed through the smaller door at the top (b).

Fig. 1.
l
z
b
b
c
g
p
n
q
b
t
s
Fig. 3.
a
b
d
c
Fig. 2.
h
g
e
h
A
i
e
d
d
h
e
h
g
Fig. 4.
m
n
n
o
c
m
a
f
a
A
s
o
g
a
m
h
a
m
n
m

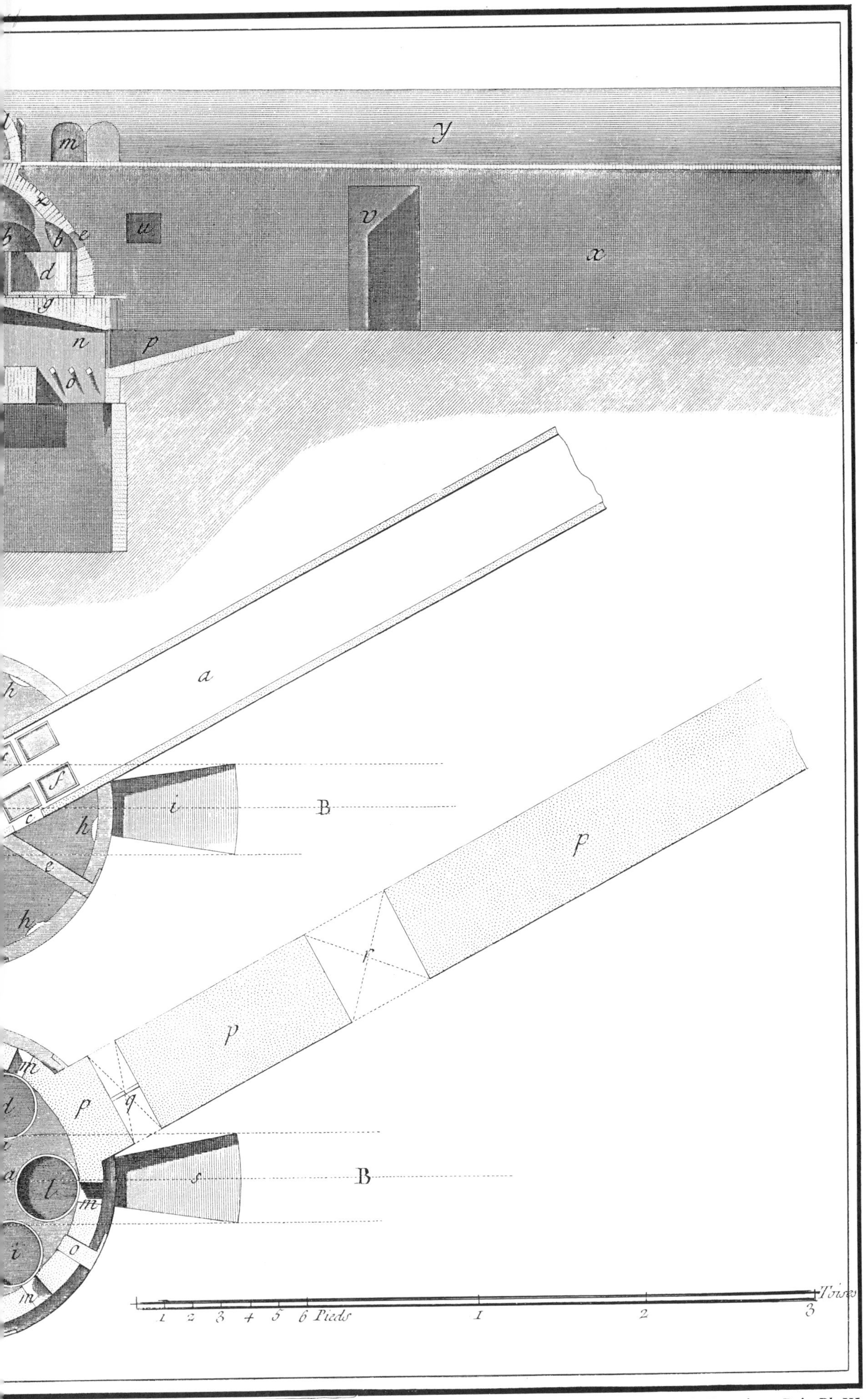
y
m
v
u
x
d
g
n
p
a
h
f
c
i
B
e
p
r
p
q
s
l
o
i
1 2 3 4 5 6 Pieds
1
2
3
Toises

Plate 211 The Glass Furnace II

Care in building the furnace and in making the pots was extremely important. Under the intense heat flaws were very likely to develop, and when, as often happened, a furnace collapsed, or a pot cracked and broke, valuable materials and working time would be lost. Even with the best of luck, molten glass corroded the pots after some months of use, and a manufacturer could not count on a furnace to serve him more than a year or two before it would be burnt out. Then it would have to be torn down and rebuilt.

The foundations (l, g) are of baked bricks, the arches and vaulting (h, r) of unbaked brickwork. The furnace is lined with refractory blocks made of argil, or potter's clay, mixed with powdered fragments of burnt-out bricks and broken pots. The pots (q) are made of the same material.

In Fig. 1 a mason (a) lays argil blocks for the hearth on a flooring of packed clay (c), and mortars the joints (b). In Fig. 3 two workmen mix powdered brick and broken pots into the argil. Bricks molded of the mixture are being dried in Fig. 4. The masons in Fig. 5 overlap the brick as they must in building the furnace—the one (a) with the hammer is knocking excess mortar loose by pounding the finished joints through a board. Once the furnace is built, it must be heated gradually until a temperature is reached at which it acts as its own kiln.

x
v
y
s
r
o
u
n
q
p
g
Fig. 5.
a
b
c
d

Vol. X, Verrerie en Bois, Pl. IV.

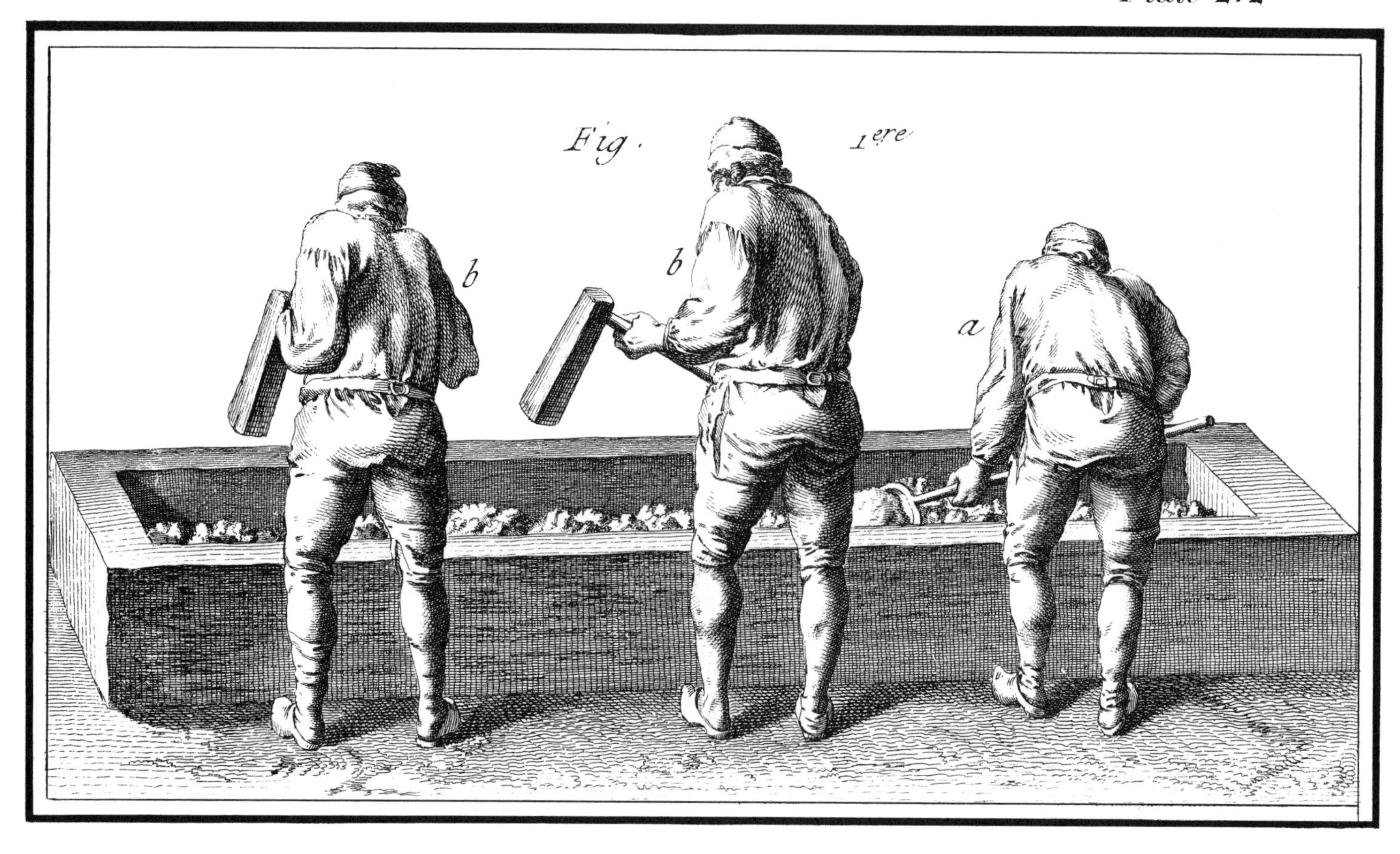

Plate 212 The Glass Furnace III

Preparing clay. In Fig. 1 argil is being pulverized by laborers whose surly backs suggest that theirs is not creative work. Casual help of this kind was hired on a daily basis, usually by an early form of the shape-up.

In Fig. 2 the clay is being moistened. Fragments of old bricks and pots are mixed in until the consistency is right for molding new bricks and pots.

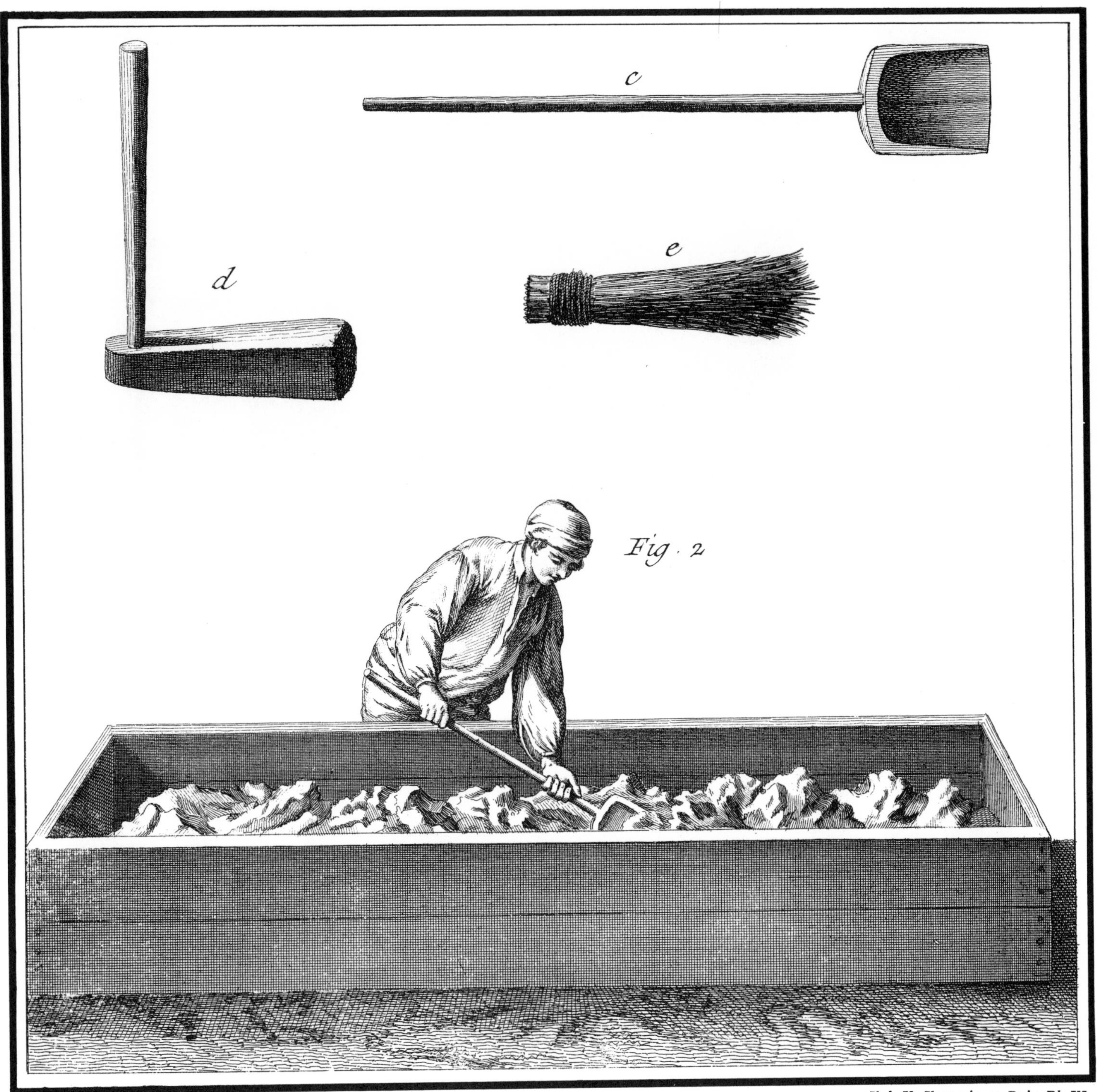

Vol. X, Verrerie en Bois, Pl. VI.

Plate 213 The Glass Furnace IV

Burnt out crucibles are broken up in mortars made of hollow tree trunks (Fig. 1). The pieces are sifted (Fig. 2), and the women in Fig. 3 tediously pick bits of glass from the fragments. The waste glass will be used as scrap.

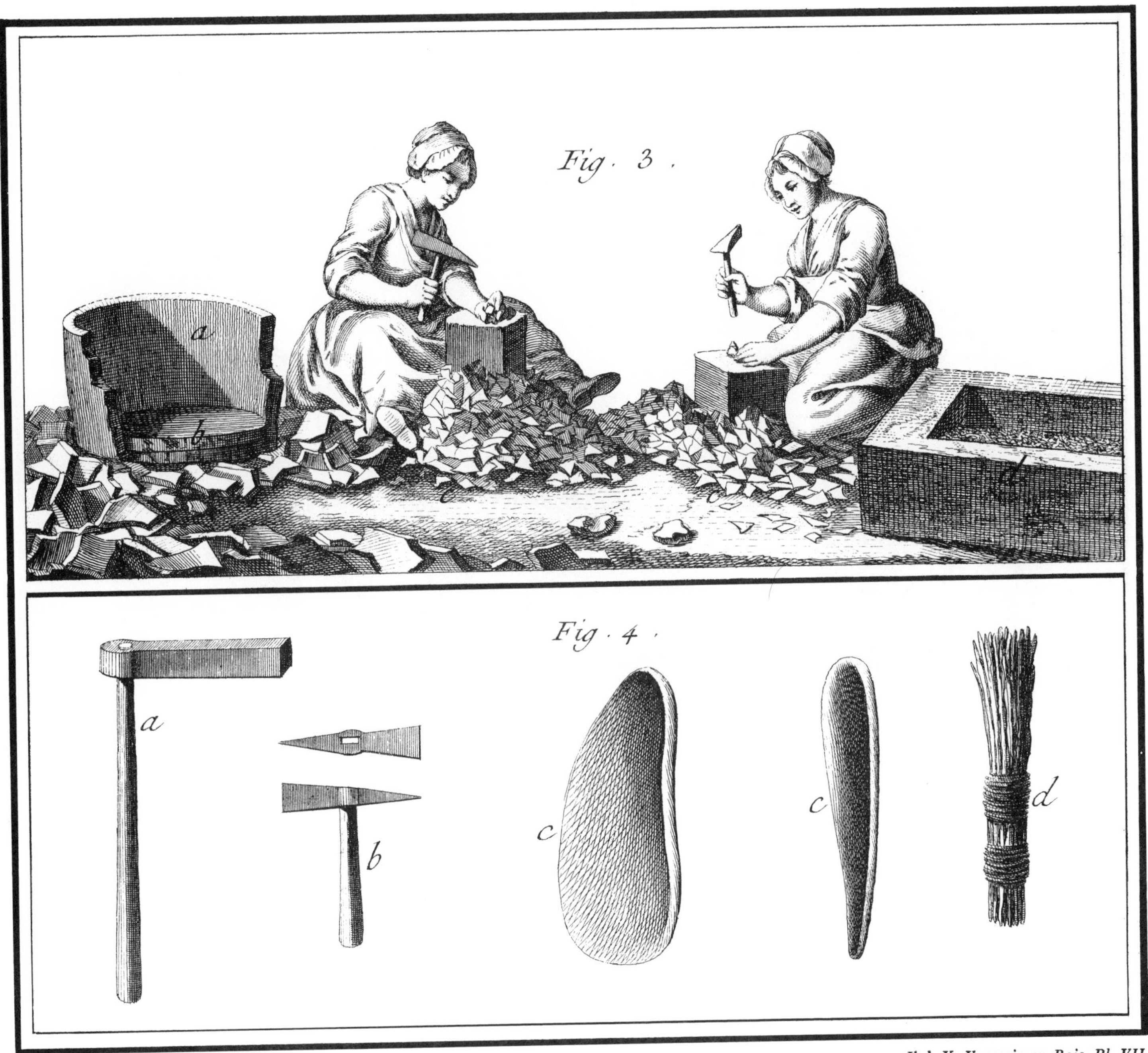

Vol. X, Verrerie en Bois, Pl. VII.

Plate 214 The Glass Furnace V

Forming the pot. In Fig. 1 a large cylinder of clay (a) is hammered out by blows on the zone indicated (b) to become the bottom. Upon this the sides are built up (Fig. 3) from rolls of clay (Fig. 2) shaped into place. Any flaw is likely to show up at the worst moment and to drown a whole run in a flood of molten glass.

Vol. X, Verrerie en Bois, Pl. VIII.

Plate 215 The Glass Furnace VI

Withdrawing a pot red-hot from the kiln after firing. The maître-tiseur *(a)—furnace-master—has charge of the construction and maintenance of the furnace, the supply of pots, and the melting of the glass composition itself.*

Plate 215 The Glass Furnace VI

Vol. X, Verrerie en Bois, Pl. X.

Plate 216 The Glass Furnace VII

Removing a burnt-out pot from the furnace (Fig. 1), and scraping the fragments from the hearth (Fig. 2).

Vol. X, Verrerie en Bois, Pl. XI.

Plate 217 The Glass Furnace VIII

Repairing a flaw which has developed in the hearth (Fig. 1). The furnace-master moistens balls of clay mixed with straw (g) with which he will reinforce the fault. In Fig. 2, a pot is being levelled after having settled to one side.

The Glass Furnace VIII

Vol. X, Verrerie en Bois, Pl. XII.

Plate 218 The Glass Furnace IX

Fig. 1 shows the tool with which the furnace-master patches the hearth. In Fig. 2 carpenters nail together a framework, a bonhomme, *against which the opening shown above will be walled up when the moment comes to close the furnace.*

Vol. X, Verrerie en Bois, Pl. XIII.

Plate 219 The Glass Furnace X

Porters carry pots from the kiln to the main furnace (Fig. 1). In Fig. 2 (opposite) the furnace door is walled up (c) with battens of fire clay against the framework built in the previous plate. This is a temporary arrangement, torn out and replaced for each operation.

Vol. X, Verrerie en Bois, Pl. XIV.

Plate 220 Making Glass I

Calcining or roasting raw materials. Glass itself was made from sand, potash or soda, and sometimes lime, mixed in varying proportions according to the qualities desired. If the product was to be white glass, a decolorant, usually manganese, would be added. The French did not make flint glass, or English crystal, though they did make white-glass goblets in imitation of Venetian crystal. The first step was to calcine the materials to a rough mass called "frit," which the furnace-man is raking from the fritting furnace or "calcar." It will next be mixed with scrap glass—"cullet"— to make the final composition for fusion in the main furnace.

Vol. X, Verrerie en Bois, Pl. XV.

Plate 221 Making Glass II

"Cullet", scrap glass to be mixed with frit, is washed, picked over, and carried to the warehouse by a pair of casual laborers.

Vol. X, Verrerie en Bois, Pl. XVI.

Plate 222 Making Glass III

In Fig. 1 above, the workmen mix frit and cullet. In Fig. 2 furnace-men charge the mixture through the opening (a) into a pot inside the furnace where it will be fused to glass.

Making Glass III

Vol. X, Verrerie en Bois, Pl. XVII.

Plate 223 Blowing Goblets I

The next two plates illustrate the making of a goblet. This was the finest product of the industry, and a skilled master-blower of goblets was its most highly paid artisan. In Fig. 1 the furnace-master stirs the batch. When he is satisfied, the gatherer (Fig. 2), collects a "parison" of molten glass on the end of a blowpipe. In Fig. 3, an assistant blower rolls it on a "marver" to an even consistency, and gives it to a master. The master then blows it (Fig. 4) in the wooden mold (c) which shapes the lower end into the bowl, and in the case of a fluted goblet imprints the pattern.

Next the stem must be joined. In Fig. 5 the master-blower forms a throat. An apprentice (Fig. 6) brings him a small ball of glass. He drops it on to the point of the throat, and (Fig. 7) works it into place, shapes it with his pincers, and clips off the excess. All this must be carried on rapidly, while the glass is still hot and soft. The blower prevents glass from sticking to his tools by spitting on them. When they do pick up bits of grit or glass, he rubs them clean on a leather-covered peg (c in Fig. 5) without losing time. Now the goblet is ready for a base, for which a lump of hot glass is prepared by the workman in Fig. 8.

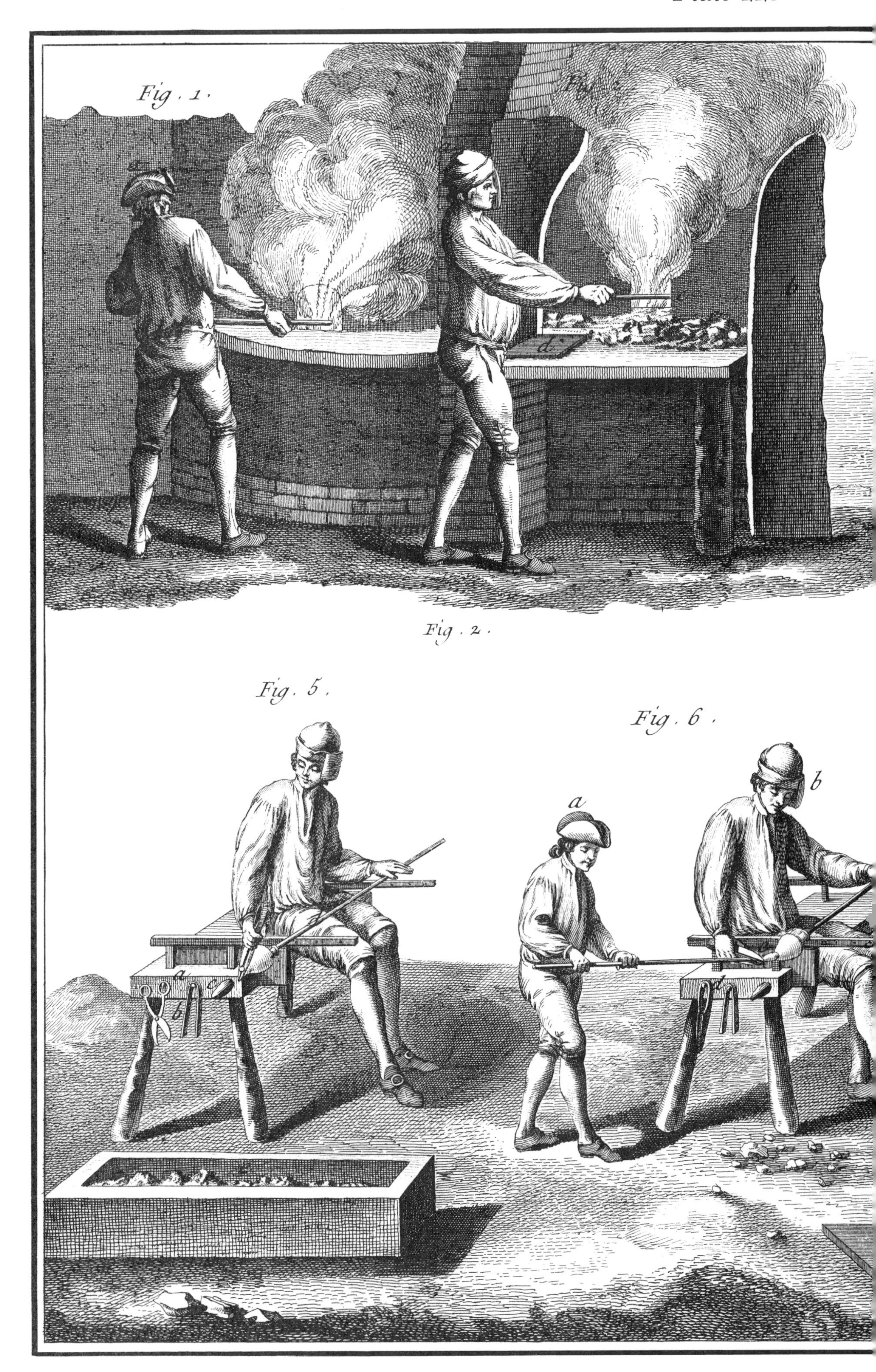
Fig. 1.
Fig. 2.
Fig. 5.
Fig. 6.
a
b
d

Blowing Goblets I

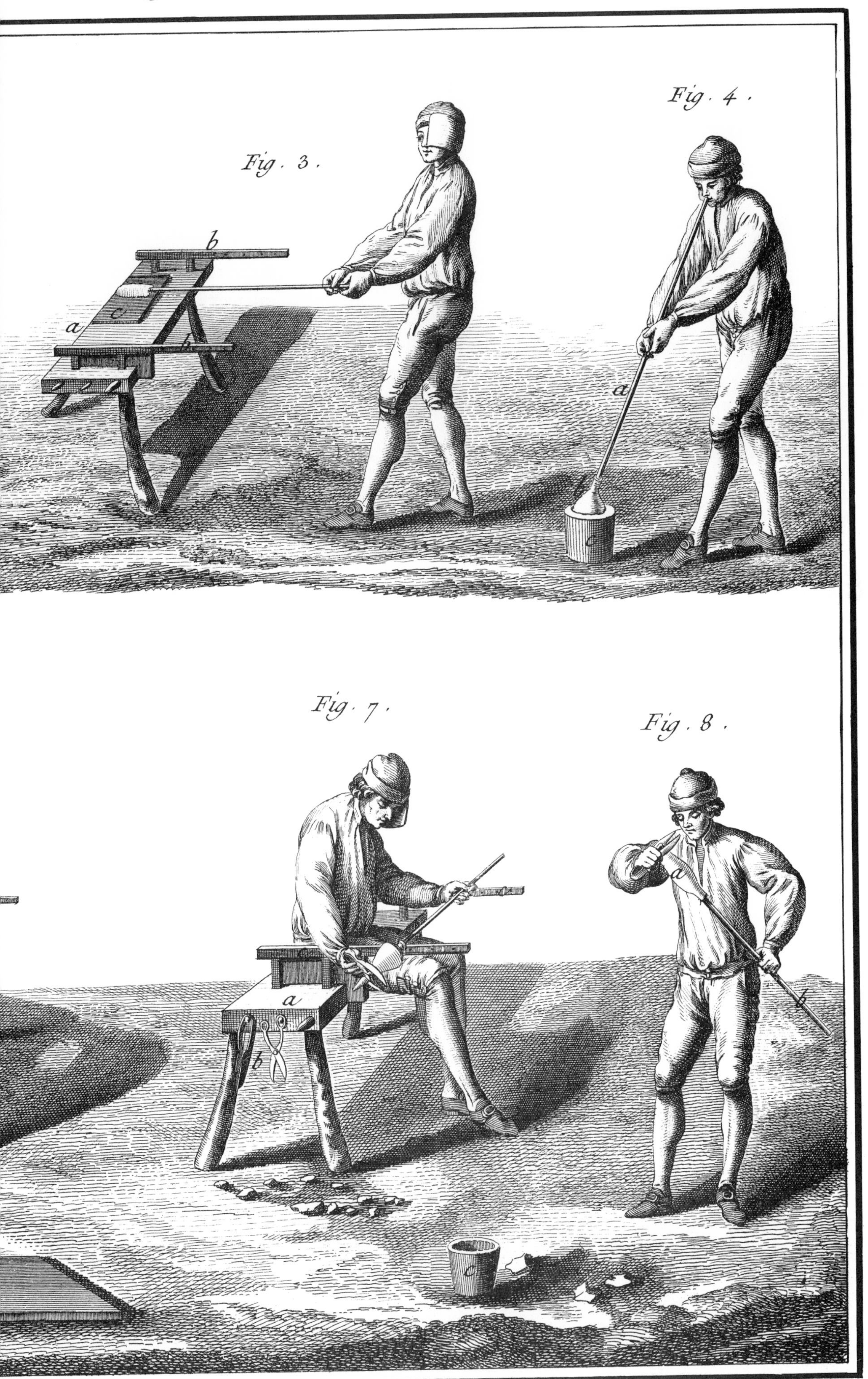

Vol. X, Verrerie en Bois, Pl. XIX.

Plate 224 Blowing Goblets II

The base is joined to the stem (Fig. 1), and the blower knocks loose the apprentice's iron rod (or "punty") with a sharp blow with the handle of his pincers. He then (Fig. 2) spreads the roll so that it becomes the disc of the base. Now he will need to handle the goblet from the bottom in order to open the mouth. In Fig. 3, he presses the base back onto his apprentice's punty, knocks loose his own pipe, and carries the glass over to the furnace. There (Fig. 4) he opens out the top by centrifugal force, softening and twirling it rapidly in the flame. All that remains is to clip the rough edges (Fig. 5), and to perfect the shape (Fig. 6) while the glass is still hot and workable. Finally, a boy lifts the finished goblets into the annealing arch over the furnace.

Fig. 1.e
d
a
c
b
f
g
Fig. 4.
a
a
e
Fig. 5.
c
d

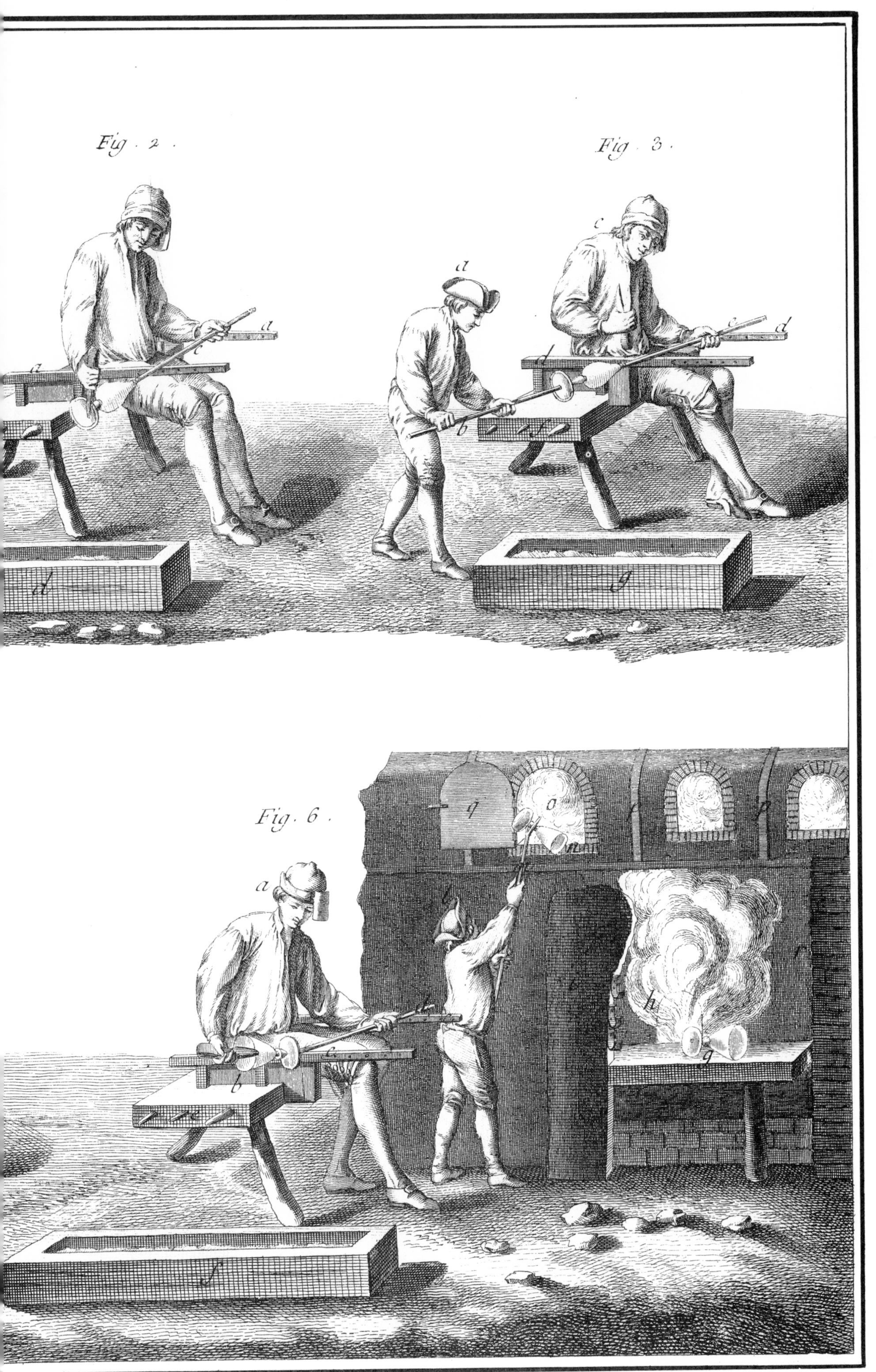

Vol. X, Verrerie en Bois, Pl. XX.

Plate 225 Blowing Goblets III

If glass cools rapidly, the difference in temperature between the outer and inner layers produces stresses which will make the piece very brittle. For this reason glass, after being blown, is passed through an annealing arch where it is cooled slowly, in some cases over a period of several days.

In this plate, one looks into the annealing oven from its outlet in the warehouse. The flue (c) is at the far end. It was through one of the two nearly invisible openings (b) that the boy in the preceding plate lifted the goblet. All the glass to be annealed is thus placed in iron trays called fraches, *which are gradually pushed from the hotter region of the furnace toward the opening. This ensures slow, even cooling. At the outlet, the*

Vol. X, Verrerie en Bois, Pl. XXII.

various kinds of glassware are unloaded by the shopmen of the sarosel *room, who keep tallies of the day's production.*

In the picture above, three shopmen carry off hampers of glasses to be packed. White-collar workers, these clerks seem a touch foppish by comparison with the men in the factory.

Plate 226 Drawn Glass

Glass is here being drawn into tubing for barometers and thermometers. In Fig. 1 the master-blower has just blown a "parison" (b), while the apprentice in Fig. 2 prepares the punty (a) with which the glass will be drawn. This is a rod spread into a disc at the end in order to get purchase in the soft glass.

In Fig. 3 the punty and the glass are dipped in water to ensure a smooth attachment, and in Fig. 4 the end of the punty is pressed into the still soft glass until the bubble closes around the disc. Then the master and the apprentice quickly change places and draw the glass out into a tube (Fig. 5). They back slowly away from each other towards either end of a row of wooden splits laid out to receive the tube when it has reached the right dimensions.

Glass-drawing requires a very sure touch, for the diameter of scientific instruments must be uniform. The tube must be drawn gently and steadily enough so that it stretches evenly, but swiftly enough so that the middle does not sag appreciably. In Fig. 6 the boys (a) are using sharp flints to cut the tubes into appropriate lengths, which are then bundled for shipping.

Fig. 1.
a
b
c
d
Fig. 2.
a
b
c
a
b
f
g
g
a
b
d
c
c
d

Vol. X, Verrerie en Bois, Pl. XXI.

Plate 227 Bottle Glass I

This is a coal-burning bottle works. Inroads on the forests had created a strong incentive for governments to encourage the shift to coal in all possible branches of industry. England was the pioneer. There the needs of the navy came first, and there the early iron industry had stripped the hillsides barer than in Europe. A proclamation of 1615 forbade English glass manufacturers to burn wood. Perforce, therefore, they learned to burn coal before iron masters did.

Similar pressure was resisted by French manufacturers and laborers, who were alike in their conservative instincts. Coal is a dirty fuel. The fumes discolored the glass and choked the artisans. To produce clear glass with coal would have required using covered pots and redesigning the factories. The typical French shop was rectangular.

Vol. X, Verrerie en Bouteilles, Pl. I.

An English factory was conical, built as a single great chimney through which a strong draft carried off the fumes. The difference is illustrated in Plate 234.

With plenty of forests still standing, French industrialists refused to make such changes in the interest of a distant future. Heavy bottles and carboys, in which discoloration did not matter, were the only articles regularly produced in coal-burning factories. The plant shown here was a "royal manufactory" at Sèvres, near Paris. The French government, ever hopeful of enticing French enterprise into more public-spirited channels, had granted it special privileges as a sort of pilot plant.

Making glass has always been a popular spectator industry, and patronage of industrial progress was a fad amongst the nobility. At left in Fig. 1, a gentleman of the court (Versailles was not far away), is conducting a great lady through the factory, perhaps explaining all the advantages of coal, though at a discreet distance from the fumes.

Plate 228 Bottle Glass II

The workman in Fig. 1 cools his blowpipe in a tub of water, and in Fig. 2 he rolls a lump of glass. A bottle-mold (e) is sunk in the ground at his feet. Notice his costume. It is different from that of the Lorraine glassworkers. This smock was the uniform for glass-workers in most parts of France. As a rule, they wore nothing else, for the heat in the shop was intense. Those whose jobs did not bring them near the furnace dispensed with the protection of the smock and worked practically naked, like bakers.

Vol. X, Verrerie en Bouteilles, Pl. II.

Plate 229 Bottle Glass III

In Fig. 1 the bottle gets a neck. The pipes (d) are used to support the blowing pipe during the blowing. The tub (f) serves to cool the equipment. In Fig. 2, right, the lump of glass is being blown into an egg-shaped parison.

Vol. X, Verrerie en Bouteilles, Pl. III.

Plate 230 Bottle Glass IV

The next stage is to roll the bottle on a marver (Fig. 1), and then shape it in a mold (Fig. 2).

Vol. X, Verrerie en Bouteilles, Pl. IV.

Plate 231 Bottle Glass V

A bottom is pressed into the bottle (Fig. 1), which is then rolled into shape (Fig. 2) and transferred from the blowpipe to a solid punty rod attached at the center of the base (Fig. 3). This will permit work on the neck.

Vol. X, Verrerie en Bouteilles, Pl. V.

Plate 232 Bottle Glass VI

Finally, a rim is threaded onto the neck. The workman drops a ribbon of hot glass from the small pipe (c), and shapes it into place (Fig. 2) with pincers (Fig. 3).

Vol. X, Verrerie en Bouteilles, Pl. VI.

Plate 233 Bottle Glass VII

Now that the bottle has been finished, it is placed in the annealing oven by a workman, out of whose skeptical and sardonic eyes looks the unchanging Parisian. He will pull loose his punty, and this will leave a rough spot at the base of the bottle. Hand-blown bottles or glasses still exhibit this imperfection.

In Fig. 2 the workman cleans scraps of glass off his blowing pipe. His hammer (Fig 3) is for breaking scrap. The tool at the right (Fig. 4) presses bottoms into bottles.

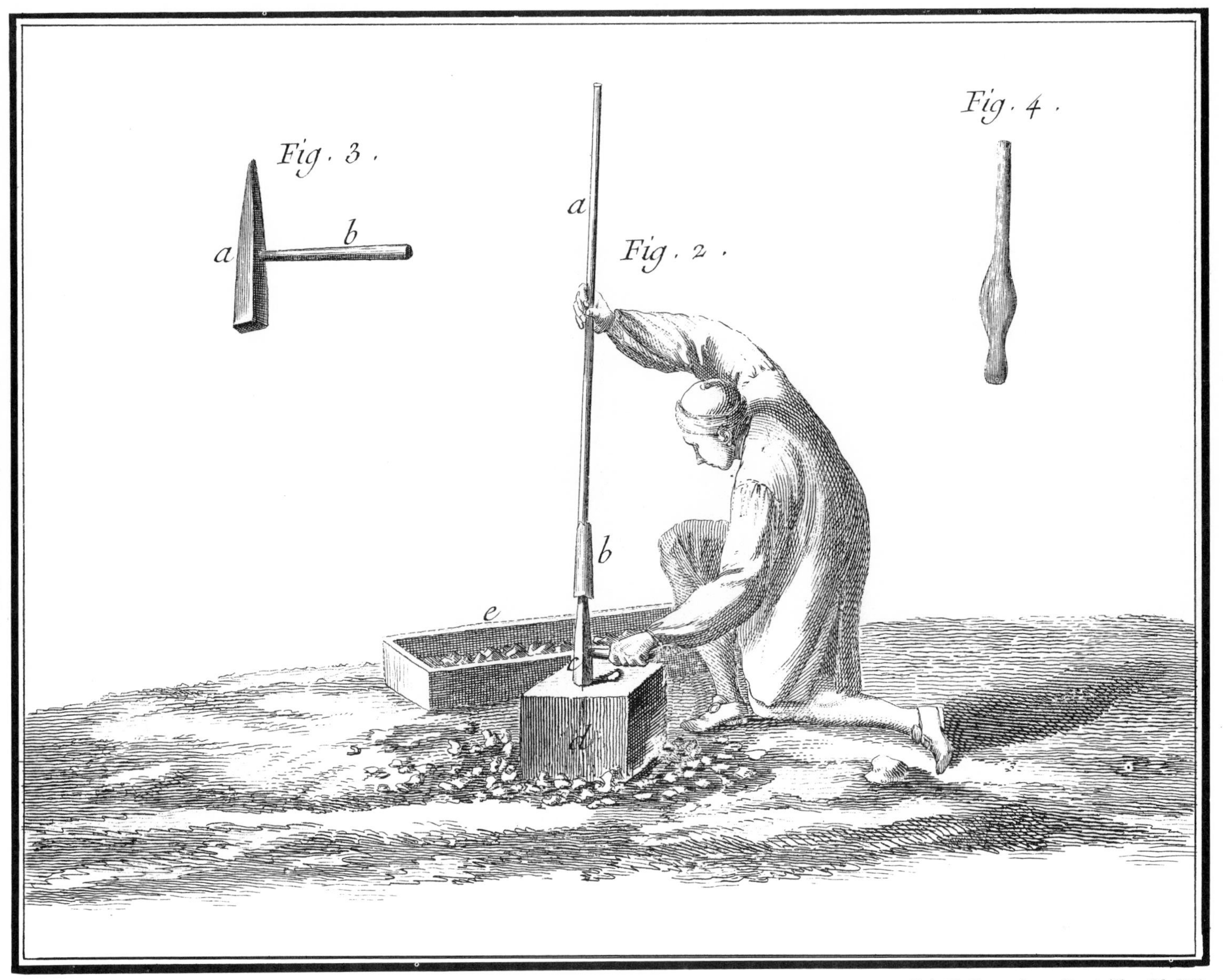

Vol. X, Verrerie en Bouteilles, Pl. VII.

Plate 234 An English Glass Works

It is clear from this picture of a coal-burning English glass-works how different was the arrangement from that of a French factory, and what a revolution would have been required to adopt English methods.

An English Glass Works

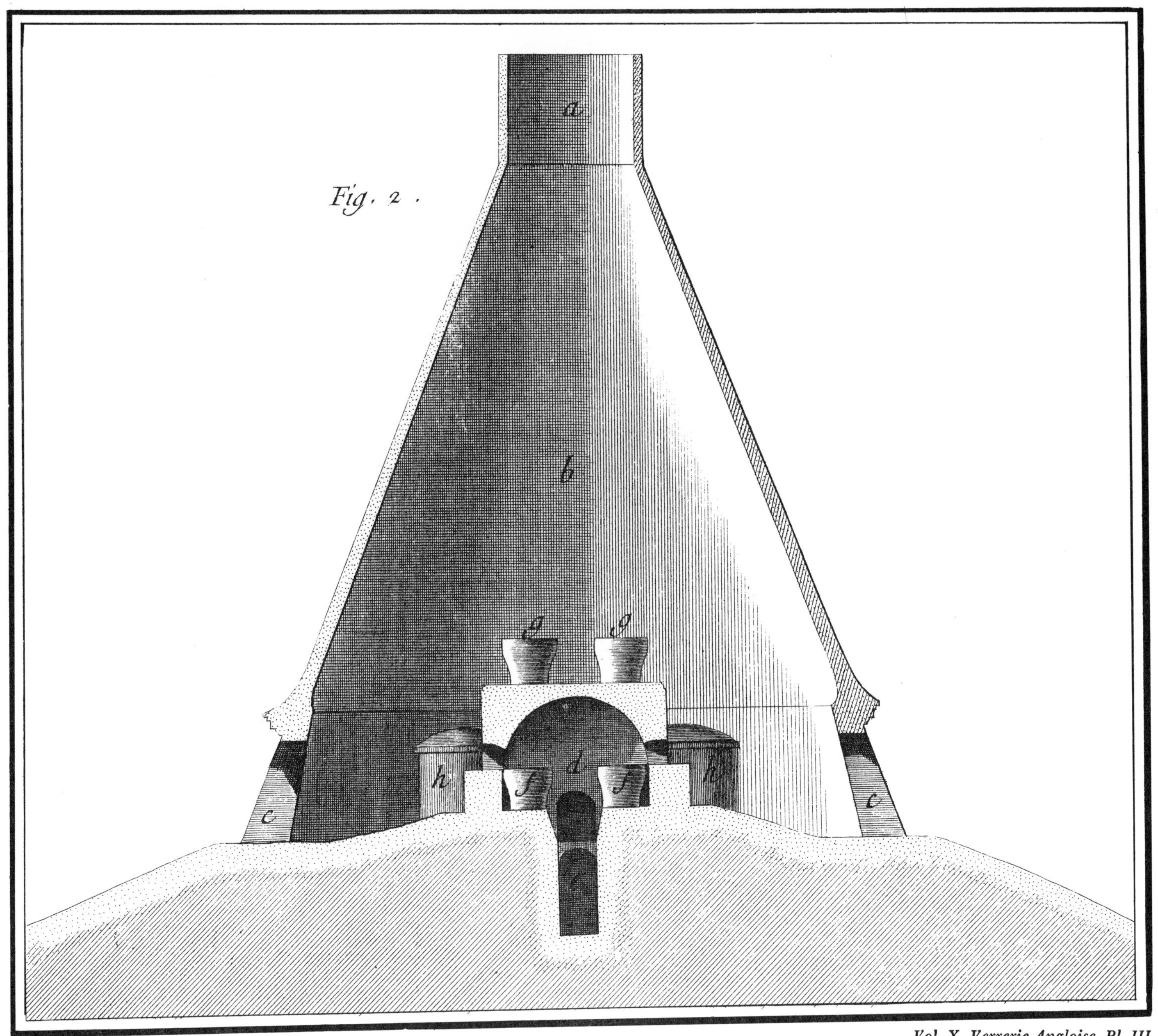

Vol. X, Verrerie Angloise, Pl. III.

Plate 235 Crown Glass I

The following plates illustrate the production of crown glass. In the 20th century, "crown glass" usually means glass of a certain chemical composition, a soda-lime glass as opposed to flint or lead glass. In original usage, however, the term defined a method of manufacture. It meant glass which had been blown into a bubble, pricked at one end, whirled flat into a disc, and used as we use plate glass. Each "table" of crown glass (Fig. 2) bore the distinguishing characteristic of a "bulls-eye", which, like the rough imperfection in the base of hand-blown bottles, was the mark left in the glass when the blower detached his punty rod.

The great advantage of crown glass was that it presented a brilliant surface unmarred by the marking and dulling inevitable in rolled or poured glass. But despite its beauty, crown glass has become obsolete. Casting or pouring plate glass (see Plates 258-275) lessened the manufacturer's dependence on highly specialized skills and produced much larger sheets for use in mirrors or fine windows. For panes of ordinary quality, crown glass was supplanted by broad glass (see Plates 249-254). The crown glass process was expensive and wasteful. Cutting the round table into rectangular sheets required heavy trimming, and besides this the bull's-eye had to be eliminated. (It is ironic that the bull's-eye, rejected as a defect when the process was in use, should have become sought after and even manufactured to lend a cachet of antiquity to shops spelled as "shoppes.")

Normandy was the center of the manufacture of crown glass. In that province there persisted into the 18th century the peculiar privilege which gave the master-blowers an aristocratic status. They are nowhere called maîtres-souffleurs *in the text accompanying these plates in the* Encyclopedia. *Instead, they are* gentilhommes-verriers*—gentlemen-glassworkers. Indeed, the profession of blowing crown glass was restricted to four families, who had enjoyed this monopoly since the 14th century. They were the Bongars, the Brossards, the Caquerays, and the Le Vaillants. As will appear, their jealously guarded art was a spectacular one.*

This plate gives a view of a crown glass factory, set in the forests which provided it with fuel and designed in the style characteristic of Norman architecture. The furnace is in the great hall (a). There are no chimneys. Smoke pours out of the windows near the ridgepole. When a run is on, a passerby might think the whole factory about to go up in flames. The entrance is to the right. An arbor (d) has been planted under which the blowers may rest and refresh themselves. The wing at the left (e) contains warehouse and workshops for making pots and incidental equipment. The pool (f) serves for cleaning tools, some of which (g) are lying beside it.

At the extreme right is a shipping container packed with circular tables of glass, each four feet in diameter. Fig. 3 gives a more detailed picture.

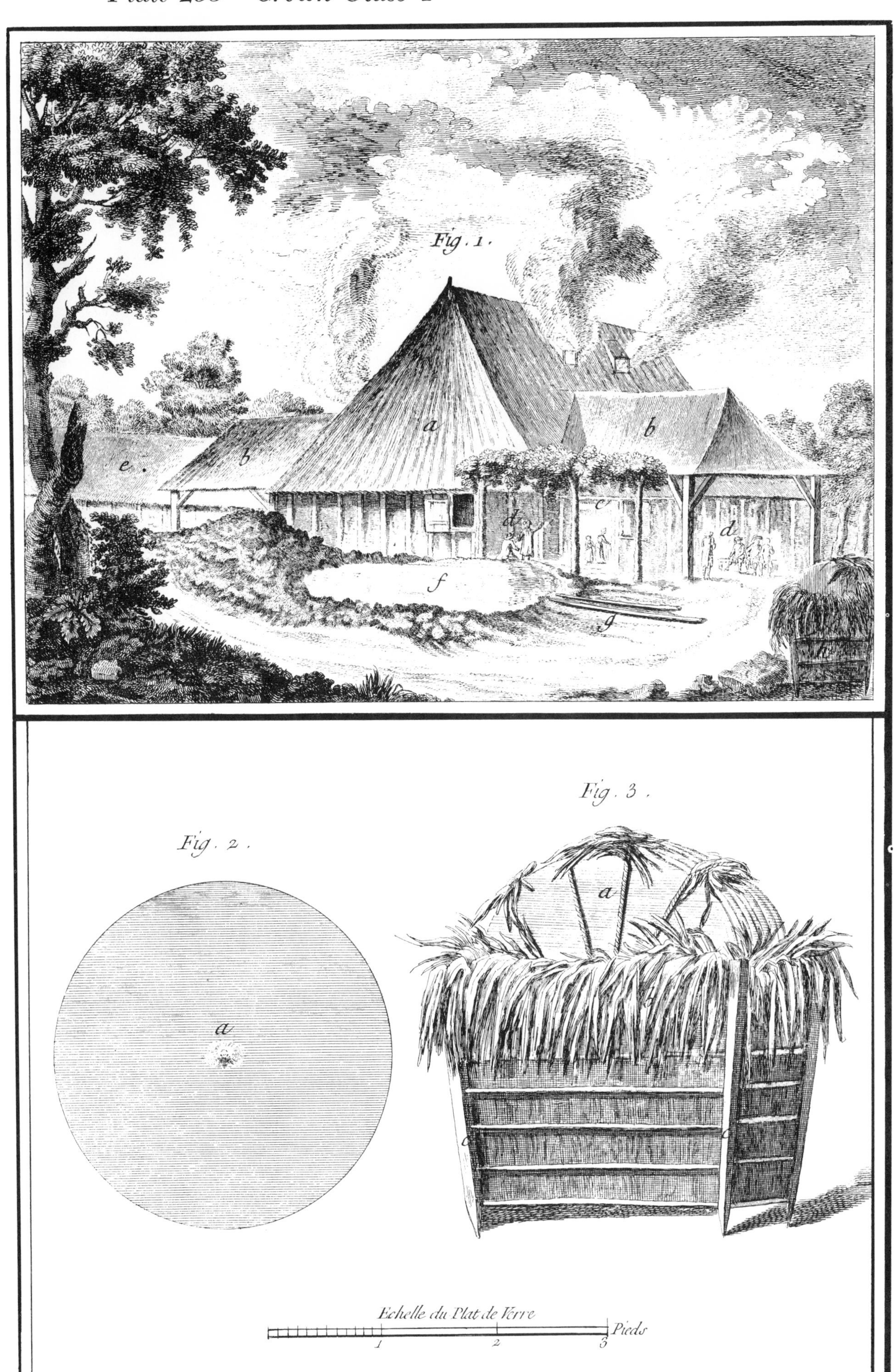
Fig. 1.
a
b
b
c
d
d
e
f
g
Fig. 2.
a
Fig. 3.
a
Echelle du Plat de Verre
Pieds
1
2
3

Plate 236 Crown Glass II

The technique of crown glass was difficult in practise though simple in principle. The tools (g) were much larger than their counterparts used for hollow-ware, and the whole operation was on a grander scale. The furnace (b) occupied the center of the shop. Kilns for the pots (c), an oven (d) for frit, and annealing ovens (t) were subsidiary to the main furnace, which was fed below (e). Wood was dried in storage over the rafters (h and i).

At the right, an assistant (o) pre-heats the end of a blowing pipe before making a gather. A gentleman-blower (p), his face shielded against the heat, lengthens a bubble—the bosse*—by rotating it rapidly and cooling it over water. A second (q) rolls it on the marver; a third (r) blows it wide; and a fourth (s) spins it open. All these steps will be illustrated in detail in the plates that follow.*

a
a
h
h
f
c
d
b
c
a
g
g
i
n
e

Vol. X, Verrerie en Bois, grande verrerie, Pl. II.

Plate 237 Crown Glass III

Fig. 1 is a lengthwise section of the furnace. The toise *of the scale at the bottom is approximately two yards—6.39 English feet to be exact. Openings (b) give access to six pots (c), of which three are shown. Air is admitted through openings (d and f), and the passages (g) lead to apertures (e) through which the furnace is stoked. At either side are ovens for roasting frit (h). Below in Fig. 2 a cross-section shows the arrangement of the combustion chamber (g). At either side are tools, with the sizes indicated.*

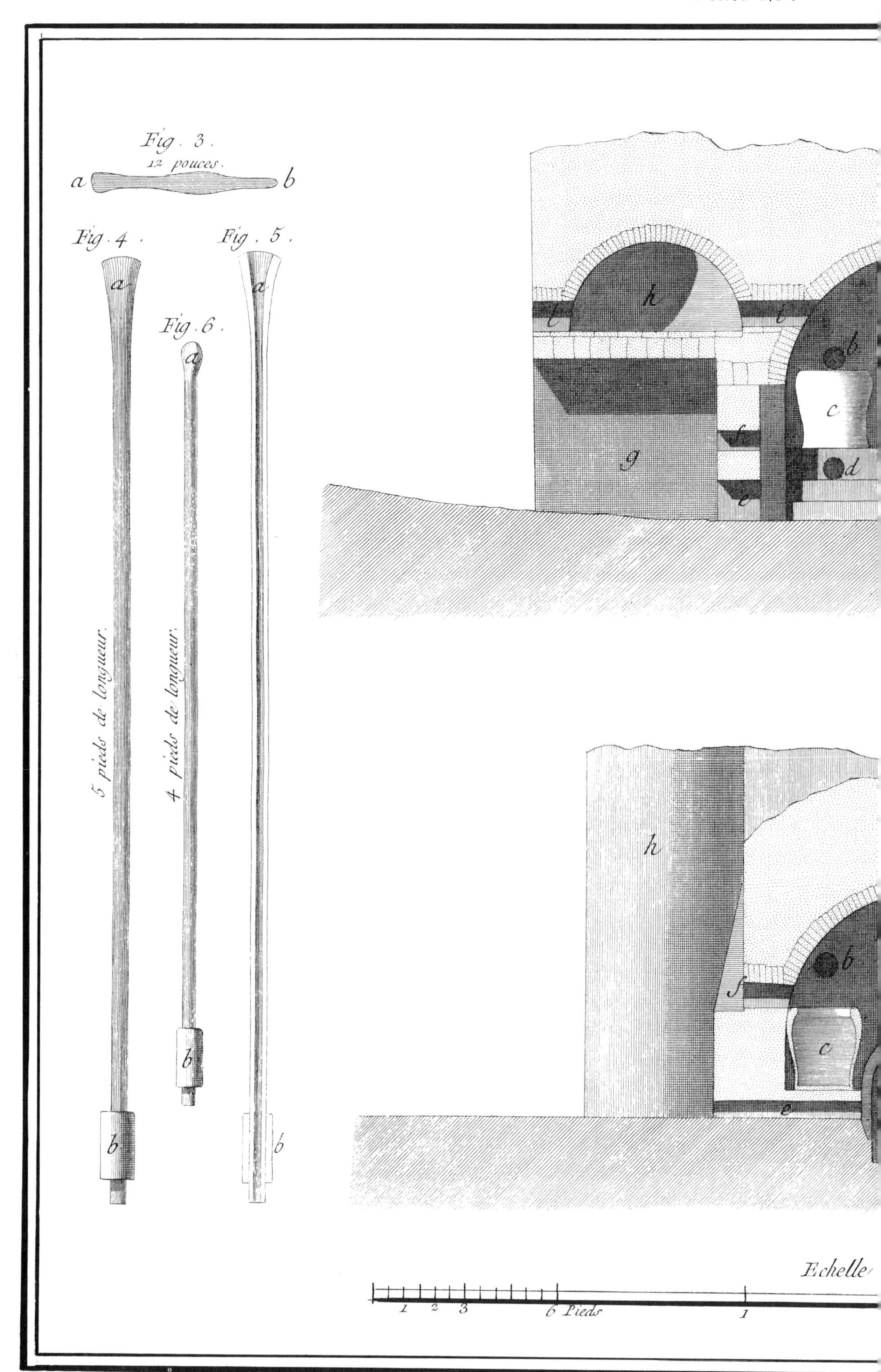
Fig. 3.
12 pouces.
a
b
Fig. 4.
Fig. 5.
Fig. 6.
5 pieds de longueur.
4 pieds de longueur.
h
i
b
c
d
e
f
g
Echelle
1 2 3
6 Pieds
1

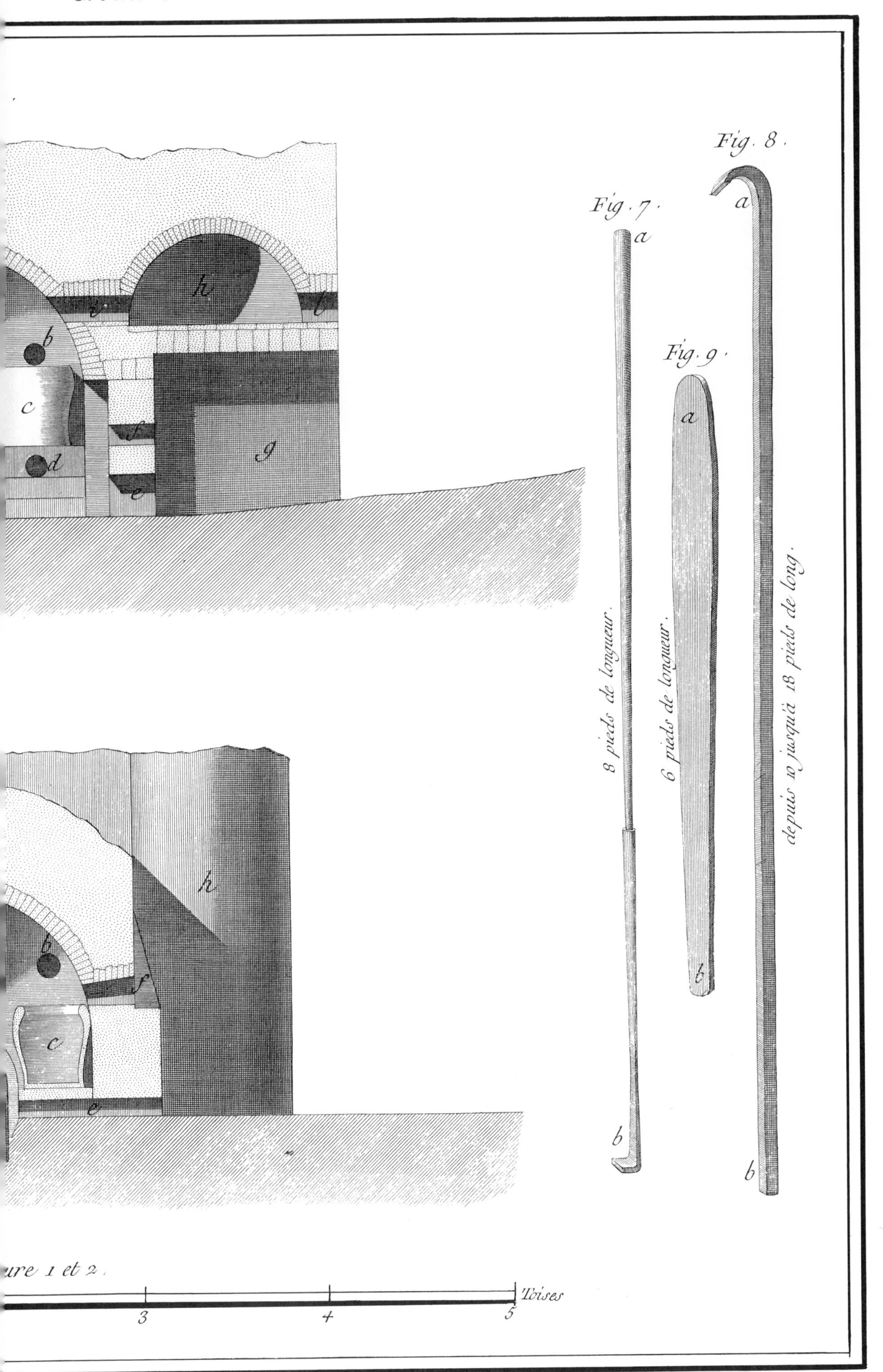
Fig. 7.
Fig. 8.
Fig. 9.
8 pieds de longueur.
6 pieds de longueur.
depuis 10 jusqu'à 18 pieds de long.
ure 1 et 2.
Toises
3
4
5

Plate 238 Crown Glass IV

The artisan must take great care in making a gather (Fig. 1). The slightest bit of grit or the smallest air bubble will spoil the product. He builds his gather by spinning his pipe in successive dips. Unless the glass is of uniform density and the bosse *perfectly symmetrical, the final table of crown glass will be warped. In Fig. 2, the blower lengthens his* bosse *to ready it for blowing.*

Vol. X, Verrerie en Bois, grande verrerie, Pl. VIII.

Plate 239 Crown Glass V

The bosse *has to be reheated between each step. In Fig. 1 the blower rolls his first heat, and in Fig. 2 he begins to blow the glass after the second heat.*

Vol. X, Verrerie en Bois, grande verrerie, Pl. IX.

Plate 240 Crown Glass VI

In Fig. 1 glass is being rolled after the third heating. After this the blower takes the bosse *to a blowing trough where he starts a neck (Fig. 2).*

Vol. X, Verrerie en Bois, grande verrerie, Pl. X.

Plate 241 Crown Glass VII

In Fig. 1 an "eye" is rolled into the lower end of the bosse. *The blowing continues in Fig. 2. Later on, this thickened "eye" will serve as the point of attachment for the punty rod on which the disc is spun flat.*

Vol. X, Verrerie en Bois, grande verrerie, Pl. XI.

Plate 242 Crown Glass VIII

In Fig. 1 the blower reheats the globe of glass, while spinning it rapidly. This flattens the side toward the fire. In Fig. 2, the blower returns to the blowing trough, where he makes a nick in the collar of the bosse. *This will allow him to remove the pipe.*

Vol. X, Verrerie en Bois, grande verrerie, Pl. XII.

Plate 243 Crown Glass IX

In Fig. 1, the blower is poised to knock the pipe loose from the neck of the bosse. *He strikes the pipe a sharp blow with the butt of his cutter, and it comes away where he has nicked the glass, leaving the neck open.*

Now he turns the bosse *on the stand and affixes a punty, a solid iron rod, to the center of the flat side where he has just dropped a small gathering of hot glass. This is applied at the eye which he had prepared (see Pl. 241) to give it purchase, and which will be called a "bull's-eye" or the "crown" when the punty is detached.*

Vol. X, Verrerie en Bois, grande verrerie, Pl. XIII.

Plate 244 Crown Glass X

It is now the turn of the opened neck to be softened by the flame. Afterwards, a boy (Fig. 2) works a specially shaped board (c) around the inside of the rim to widen it.

Vol. X, Verrerie en Bois, grande verrerie, Pl. XIV.

Plate 245 Crown Glass XI

Here is the climax of the process. The blower (Fig. 1) reheats the opened bosse, *spinning his rod rapidly. As the glass softens, it flares out, first into a shallow bell (Fig. 1) and then into a gleaming flat table of glass.*

The gentleman-blower, the aristocrat of the industry, needs strength and address. The table is heavy. If he falters in twirling it the product will be twisted or lopsided. When the glass is cool enough to be hard, it is laid into a depression (Fig. 2 d, right) prepared in the sand.

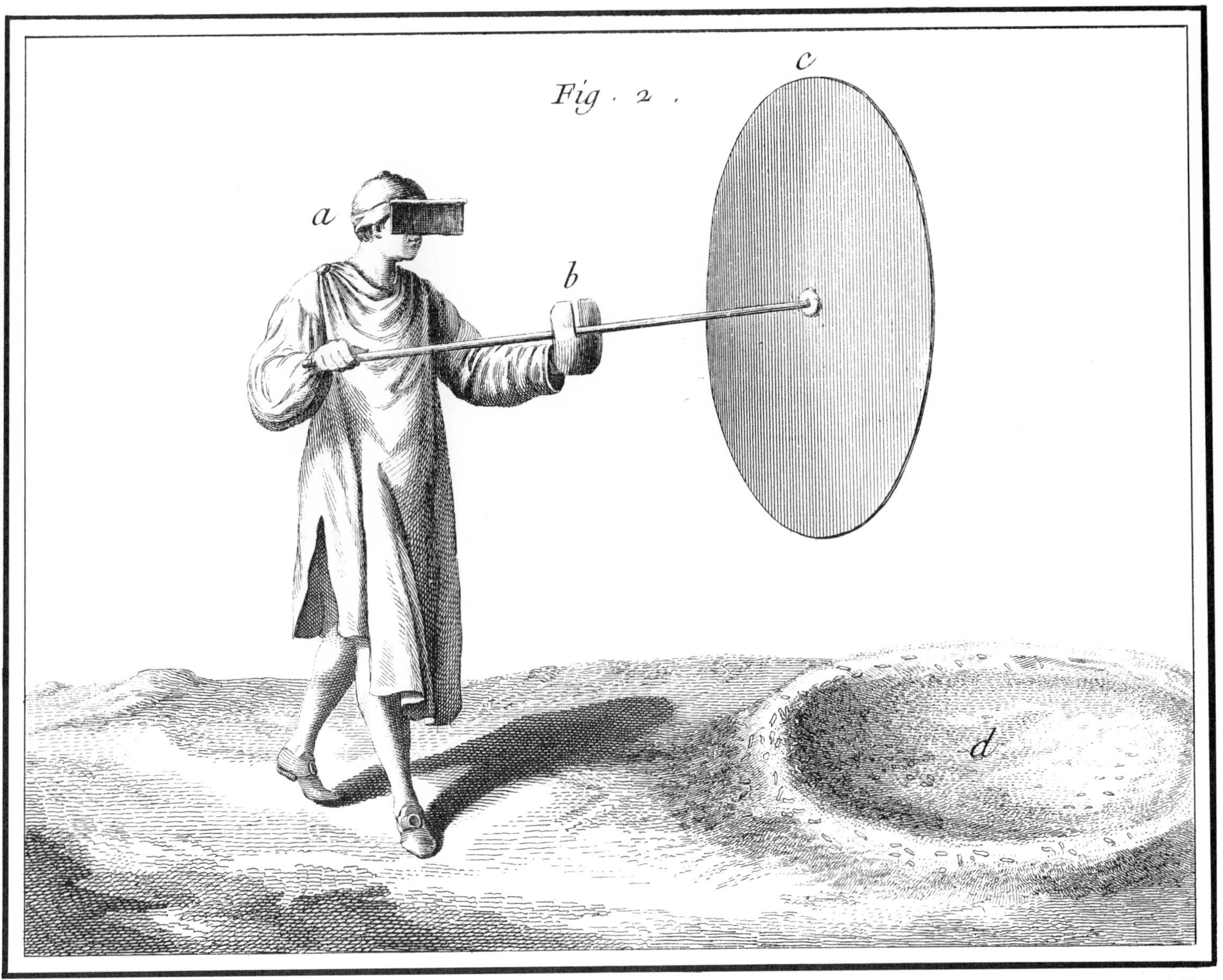

Vol. X, Verrerie en Bois, grande verrerie, Pl. XV.

Plate 246 Crown Glass XII

The punty, finally, is removed (Fig. 1) before the bull's-eye is cold, and the completed table (Fig. 2) is slid into an annealing oven.

Vol. X, Verrerie en Bois, grande verrerie, Pl. XVI.

Plate 247 Crown Glass XIII

Pots for crown glass had to be very thick to withstand the corrosive action of the melt as long as possible. Drawing a pot red-hot from the kiln and carrying it to the glass furnace was an awkward and dangerous job.

Plate 248 Crown Glass XIV

In Fig. 1, the pot, still red-hot, is being lifted up to the siege by the lever. It is steadied and will be guided into place by the grappling rod in the hands of the porters at the left. The size of the tools gives an idea of the weight of the pot.

The woman in Fig. 2 is a sweeper who rakes up broken glass in the shop. This glass is used for scrap. The pan (c) sunk in the floor is used in the control of each batch. During the run samples are drawn off by the furnace master, who allows them to cool in the depression.

Vol. X, Verrerie en Bois, grande verrerie, Pl. XVII.

g
g
g
g
g
q
f
n
b
p
a
Fig. 2.
b
c

Vol. X, Verrerie en Bois, grande verrerie, Pl. XVIII.

Vol. IV, Glaces soufflées, Pl. XXXIV.

Plate 249 Broad Glass I

In the glass industry of Lorraine and the northeast, a different technique was used for opening bubbles of blown glass into sheets. The first steps were similar to the crown glass process, but then instead of opening the glass into a bell, the blower (f) rotated the glass vertically into a hollow cylinder. This was then slit down the side and pressed into sheet glass. This was the method of the glass industry of Bohemia and Germany, of which Alsace and Lorraine were, in certain ways, technological outposts. For this reason broad glass was also called "German sheet."

The typically French smock, it will be noticed, was not worn here. Instead, the artisans are dressed like those of the small-ware shop (Plates 209-226), which was also in Lorraine.

Plate 250 Broad Glass II

Vol. IV, Glaces soufflées, Pl. XXXVI.

Plate 250 Broad Glass II

An artisan (a) takes the glass on a punty at the opened end, while the blower (b) knocks it off his pipe. A third artisan (d) spreads the neck so that the cylinder is of roughly constant diameter, after which it is carried up on a sort of platform so that the slitter (g) can cut it with his shears. After slitting, the glass is unrolled as a sheet on the table at the rear (h).

Vol. IV, Glaces souflées, Pl. XXXVII.

Plate 251 Broad Glass III

Annealing broad glass.

Plate 252 Broad Glass IV

Here are the tools used in the manufacture of broad glass.

The toise *of the scale at the bottom is approximately six feet. It is evident that the sheets of glass must be fairly small—about three feet by four.*

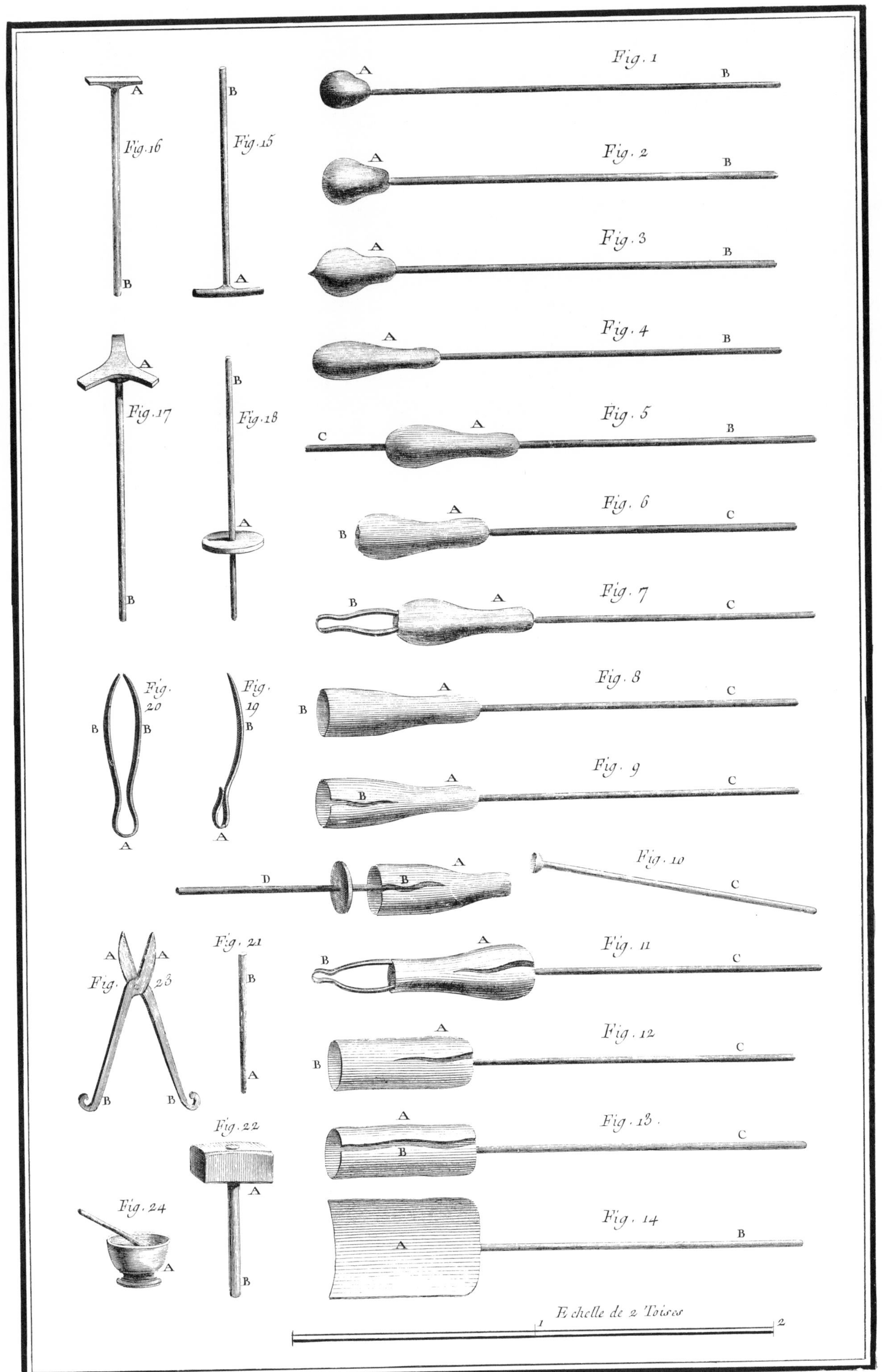
Fig. 1
Fig. 2
Fig. 3
Fig. 4
Fig. 5
Fig. 6
Fig. 7
Fig. 8
Fig. 9
Fig. 10
Fig. 11
Fig. 12
Fig. 13
Fig. 14
Fig. 15
Fig. 16
Fig. 17
Fig. 18
Fig. 19
Fig. 20
Fig. 21
Fig. 22
Fig. 23
Fig. 24
Echelle de 2 Toises
1
2

Vol. IV, Glaces, Pl. XXXIX.

Plate 253 Broad Glass V

Since broad glass was unrolled on a flat table while still soft, its surface was dulled and scarred. For this reason, and also because the sheets were never uniformly thick, it had to be evened and polished by grinding. Even so, it never presented the brilliant surface which was the great characteristic of crown glass. Despite the expense of the extra step of polishing, however, the broad glass process required less dexterity than did crown glass, and since the product was rectangular and contained no bull's-eye, it could be used in larger panes.

This is a grinding shop. Glass is made to grind glass. Large sheets are plastered to the tables, while smaller sheets are fixed either to slabs of limestone fitted with handles (b, c, and d), or to the spokes of a wheel (a). Emery and wet sand are spread on the surfaces, which then grind each other smooth. Unground sheets are stored in the crude rack (e) at the back. No doubt the windows of the shop itself are glazed with factory rejects.

Vol. IV, Glaces, Pl. XXX XII.

Plate 254 Broad Glass VI

Finally, broad glass had to be polished. The surface was furbished with felt buffers affixed to jointed ribs which maintained a constant pressure. Bits of grit or sand were brushed off with a silken brush. The wall mirror on the workbench in the background is the sort that was made of glass of this quality. Larger and finer mirrors, however, like those used in elegant salons or boudoirs, could be made only of the best grade of plate glass.

Plates 255, 256 Mirrors I, II

Diderot seems to have had two engravings illustrating the silvering of mirrors, and he included both, perhaps because one is clear and the other lifelike—though that is not the *mot juste*, perhaps, since like other artisans who handled mercury, most of them would be poisoned quite young.

This was the method of silvering mirrors between the late Middle Ages, before which time most mirrors were simply polished metals, and the 19th century, when the backing of glass with silver made silvering an accurate term. Like other techniques of working glass, the art of silvering was introduced into France by certain Venetian workmen from Murano who were persuaded to immigrate by agents of Colbert—"abducted by Colbert," the Venetians would have said.

The "silvering" is actually speculum metal, an amalgam of tin and mercury. (Since the two plates represent the same process in two different shops, they can be explained together, the letter referring to the upper vignette, and the numeral to the lower). A sheet of tinfoil is laid out absolutely free of creases or wrinkles. On it is poured a little mercury (b) which is rubbed over the surface. The loosened scum of dust and impurities is removed (2 at the right). Then the tin is covered with perhaps a quarter of an inch of mercury, just enough to float the glass, a plate of which has meanwhile been polished to the utmost (a, 2 left). The glass is slid on to its thin cushion of liquid mercury (c, 3) and weighted solidly with rocks (f) until all the excess mercury has been squeezed out around the edges. Then the mirror in the making is moved (d, 1ere) to a tilted rack where, silver side up, it drains, dries, and hardens for 24 hours.

Vol. VIII, Miroitier metteur au teint, Pl. I, et Miroitier, Pl. I.

Plate 257 Plate Glass I

Though the French glass industry was extensive and expanding, none of the methods shown so far was indigenous. Exclusively French, however, was the technique of casting or pouring molten glass in plates, the most important innovation since the prehistoric discovery of glass itself. Considerable obscurity surrounds the circumstances. Neither the inventor nor the date is certainly known, although Barrelet[1] *does credit the discovery to Bernard Perrot, or Perrotto, the naturalized descendant of a famous family of Altarese glassworkers. It is agreed however, that plate glass was in production from 1688. The principle of the process is that glass is handled like metal. This admits of the manufacture of far larger and more uniform plates than any method dependent on blowing. And though casting plate glass required a heavy investment in plant, the compensation was an extensive substitution of unskilled labor for skilled.*

Almost all 18th century plate glass was manufactured by a single concern, the Royal Plate Glass Company, which had been founded in 1665 under the patronage of Louis XIV. It had plants in Saint-Gobain in Picardy and at Tourlaville in Normandy, and a polishing plant and warehouse outside Paris in the Faubourg Saint-Antoine. At the outbreak of the Second World War, this concern was still one of the largest in the world. Its operations are illustrated in the plates that follow.

As always, the first step is fabrication of pots and refractory brick. The porters are carrying argil clay to be mixed with water and thoroughly trodden to the consistency of thick mud. In a "royal manufactory" like this one, the chief porter might also act as a doorman on appropriate occasions, and would then wear the royal livery. But it is not likely that his dejected assistant at the rear of the barrow would be called on for ceremonial duty.

[1] La verrerie en France *by James Barrelet.*

Vol. IV, Glaces, Pl. IV.

Vol. IV, Glaces, Pl. V.

Plate 258 Plate Glass II

Pots made of damp clay (I, H) are molded inside great wooden buckets (Figs. 1, 2, and 4). Plate glass was fused in pots, but handled in smaller rectangular vessels like the one being molded in Fig. 3. These served as great ladles for pouring glass, though they are not what one ordinarily pictures as ladles. Two sizes (M, N) were required. Each poured a single sheet, the smaller being half the size of the larger.

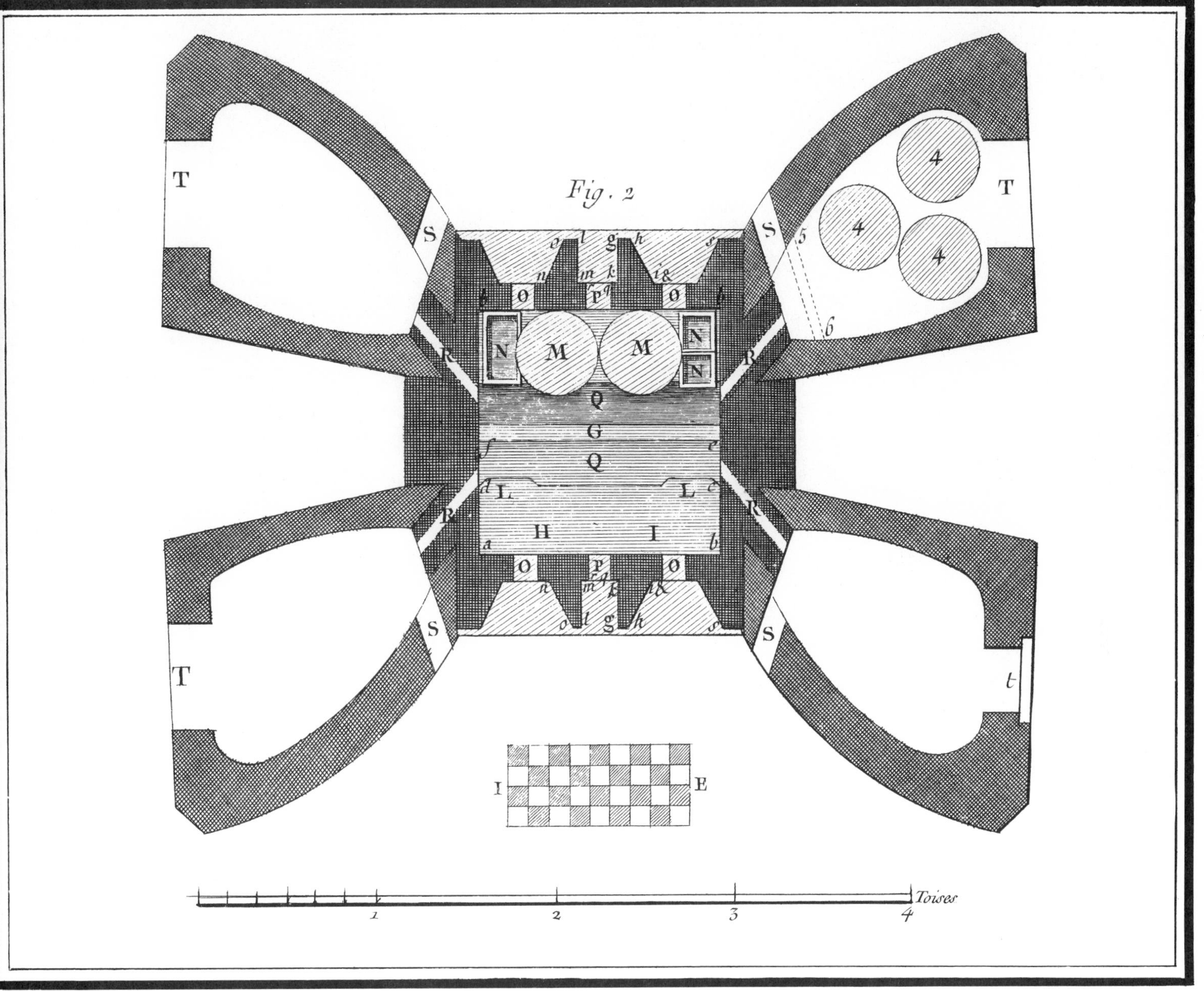

Vol. IV, Glaces, Pl. VI.

Plate 259 Plate Glass III

This is a horizontal section of the furnace, reproduced in order to clarify the purpose of the steps to follow. Molten glass had to be transferred from the melting pot to the pouring ladle inside the furnace. The respective locations for the pots (H, I) and ladles (L, L) are indicated on the hearth shown at the bottom (a, d, c, b). On the hearth shown above, two pots (M) and three ladles (N) are in place. The glass was worked through openings (o) in front of each pot. Flues (R) led waste heat to kilns placed in the corners. Three of these (T) were used for firing the pots (4) and ladles. The fourth (t) preheated frit and cullet which were transported to the main furnace as shown in the following plates.

Vol. IV, Glaces, Pl. XXVII.

Plate 260 Plate Glass IV

Placing a pot in the kiln before firing. (The reader may wish to refer back to the previous plate for the design of the furnace). Evidently the fires have not yet been lighted.

Vol. IV, Glaces, Pl. XXVIII.

Plate 261 Plate Glass V

Withdrawing the pot red-hot after firing. The carriage is designed to provide a considerable mechanical advantage. But notice the number of handles. Despite the leverage, seven men are needed to move the pot even when it is empty.

Vol. IV, Glaces, Pl. XXIX.

Plate 262 Plate Glass VI

A cutaway of the furnace, showing a pot being seated on the siege. The passage between the two sieges leads up from the fire grate below. Many 18th century tools have graphic names. The grappling hook (a) is called dent du loup*—a wolf-fang. The instrument used to scratch out the inside of the ladle in Plate 268 is called* la grand'mère*—the grandma.*

Plate 263 Plate Glass VII

Vol. IV, Glaces, Pl. XXX.

Plate 263 Plate Glass VII

After the pots are in place, a ladle is taken from the kiln (1 and 2). It is apparent how the smaller ladle being carried (4 and 5) to the furnace is grooved to allow it to be grasped by the talons of a pincer-carriage (6 and 7).

Vol. IV, Glaces, Pl. X

Plate 264 Plate Glass VIII

The raw materials, sand and pulverized limestone, are sifted and washed free of impurities. At the left a stream of water runs right through the shop.

Vol. IV, Glaces, Pl. XII.

Plate 265 Plate Glass IX

Calcining "frit."

Wheeled bins (1 and 2) bring the raw materials up to the doors of the fritting furnace. Furnace-men stir the mass with the fire-rakes at the left (3 and 4). In Fig. 1 the workman is changing a rake which is too hot to handle for a cool one.

Vol. IV, Glaces, Pl. XVIII

Plate 266 Plate Glass X

A mixture of frit and cullet is preheated in a kiln from which a workman (1) withdraws it. Transferring it to the main glass furnace is a continuous process. It is very hot and is carried around by scoopfuls (2), and thrust (3) through an opening into a pot where it is to be fused. The men return for another load (4) and await their turn (5 and 6). A furnace master supervises the progress of the melt (7). Apparently the heat is less intense than near a crown glass furnace, for the workmen are shielded by only their hat-brims. Fuel dries across the rafters.

Vol. IV, Glaces, Pl. XXXI.

Plate 267 Plate Glass XI

Inevitably, a little glass gets spilled and drips down over the sieges into the fire, where it is burnt to the consistency of heavy clay. This waste, called picadil, *is here being cleaned out through the door to the fire-pit. The workmen must cool their scrapers and dippers in the tub of water every few minutes.*

Vol. IV, Glaces, Pl. XIX.

Plate 268 Plate Glass XII

An empty ladle is withdrawn red-hot from the furnace to be cleaned out (1 and 2) before charging. It had first to be heated in order to melt the glass adhering to the sides from the previous run. The ladle is handled by means of a chariot-like device which is essentially a huge pair of pincers on wheels. The shafts are hinged scissor-fashion over the axle so that the ladle can be grasped along grooves molded into the sides. One workman (3) scrapes the inside surface, and a second (3) pours the waste into a shop-boy's dipper. (Children were regularly employed for unskilled work in most 18th century industries). After cleaning, the ladle will be returned to the furnace until the fused glass is ready to be transferred from the pots.

Vol. IV, Glaces, Pl. XX.

Plate 269 Plate Glass XIII

Impurities are skimmed from the surface of the melting pot (1). The molten glass adhering to the rod is scraped off on a slab (2), to be reclaimed for cullet. A stoker (3) adds wood to the fire.

Vol. IV, Glaces, Pl. XXI.

Plate 270 Plate Glass XIV

Transferring thoroughly fused glass from the pot to a ladle inside the furnace. A workman (1) thrusts his dipper right down to the bottom of the pot. Another (2) pours molten glass into the ladle, while a third (3) cools his dipper in a cask of water and a fourth (4) stands ready to take his place in front of the pot.

Before this step was possible, the process of fusion had to be arrested and the glass allowed to quiet down in the pot. No further fuel was added during this period, which gave the workers a respite and which was known among them, therefore, as the "ceremony." Thus to stop the glass was to faire la cérémonie—*to perform the ceremony.*

Plate 271 Plate Glass XV

Vol. IV, Glaces, Pl. XXII.

Plate 271 Plate Glass XV

A ladle of molten glass, ready to pour, is pulled out of the furnace onto a carriage. This is a tense moment. Everyone must move quickly to prevent the glass from overcooling. The furnace-master (1) himself takes a hand. The carriage will be wheeled rapidly over to a casting table in front of one of the annealing ovens in the background.

Vol. IV, Glaces, Pl. XXII

Plate 272 Plate Glass XVI

While the pourers (1 and 2) and their assistants (3 and 4) skim off the last impurities, which will be disposed of by the inevitable gamin *(5), a sling is adjusted to hoist the ladle from its carriage and swing it over the casting table. This is a rectangular slab of copper with a shallow rim, in form not unlike a billiard table open at both ends. It is mounted on wheels and can be moved laterally along a track running across the face of a battery of annealing ovens.*

Vol. IV, Glaces, Pl. XXIV.

Plate 273 Plate Glass XVII

Pouring a glass plate is the climax of the operation. It requires speed and minute attention from every workman. The glass must be poured (1 and 2), and rolled out (3 and 4) evenly. Under the copper roller it handles like dough under a rolling pin and presents a workman with the same inert resistance to his will. Assistants (5 and 6) guide the roller. Others (7 and 8) are alert to catch and pluck out any bits of dust or scale which might fall onto the still soft surface.

Behind the two rollers, another pair (9 and 10) loosen the iron strips rimming the table before the glass has hardened to them, but only when it is stiff enough not to run. The hoistman (11) gives the pourers the height they need. The master (12) keeps his eye on everything, and the porters (13) take the carriage back to the furnace to get another ladle, for which the table will be moved along to the next oven.

Vol. IV, Glaces, Pl. XXV.

Plate 274 Plate Glass XVIII

The plate is slid into the annealing oven. Annealing plate glass required ten days. This shop has sixteen annealing ovens in operation to handle the output of a single furnace.

Vol. IV, Glaces, Pl. XXVI.

Plate 275 Plate Glass XIX

Drawing the plate from the annealing oven (1-7) and carrying it away (8) to be ground and polished. The artist has probably exaggerated its transparency. As a rule the surface of plate glass was so dulled by casting and rolling that it was almost opaque before being polished.

Masonry & Carpentry

Plate 276 Masonry

In cities built of stone, the mason had the central place in the building trades which in America is occupied by the carpenter. This plate is arranged to illustrate his different tasks in the construction of some grand house: (A) hoisting stones all cut to fit; (B) mortaring joints; (C) truing a footing; and (D and E) marking stones with rule and calipers and cutting them to measure. Almost lost in the background (G) is a mason sawing a large block. Various laborers mix mortar (F) and haul sand and plaster about (I, K).

Sidewalk superintendents of present-day construction in Paris may observe scenes that are not very different—which, indeed, is one reason that construction moves more slowly in France than anywhere else in the western world. The costumes of the workers have altered more than have tools or techniques, and although the hoist (A) would hardly be seen on a big construction job, the scaffolding (B) is indistinguishable from one that the present writer saw in use for repairs being made at the French National Archives in the summer of 1955.

Vol. I, Architecture, Maçonnerie, Carrier Platrier.

Plate 277 Mining Gypsum

Gypsum for use in plasterwork was quarried in the neighborhood of Paris itself—hence "plaster of Paris." It is packed out of the mines on donkeys and calcined or burnt right at the site.

Plate 278 Tiles I

The Tuileries gardens in Paris, like the royal palace burnt by the mob in 1870, take their name from a tile factory which had stood on the site in ancient days when it was outside the city walls. Used for both flooring and roofing, tiles were a building material of great importance.

This is the ensemble of an 18th century tuilerie, *with oven (ABC) and service shed (DE) at the left, an area for sun-drying down the center, and the mixing and molding shop (F) at right. The work of the latter is shown in a foreshortened view opposite. A worker (Fig. 4) treads clay in a bin, supplying himself with water as needed from the well-hoist (n) which appears in the upper picture. The molder then presses the damp clay into oblong forms. He dusts his molds and working surface with sand to keep the clay from adhering, and hands the products over to a dryer (Fig. 2) who lays them out on the carefully swept (Fig. 1) ground to bake in the sun until almost dry.*

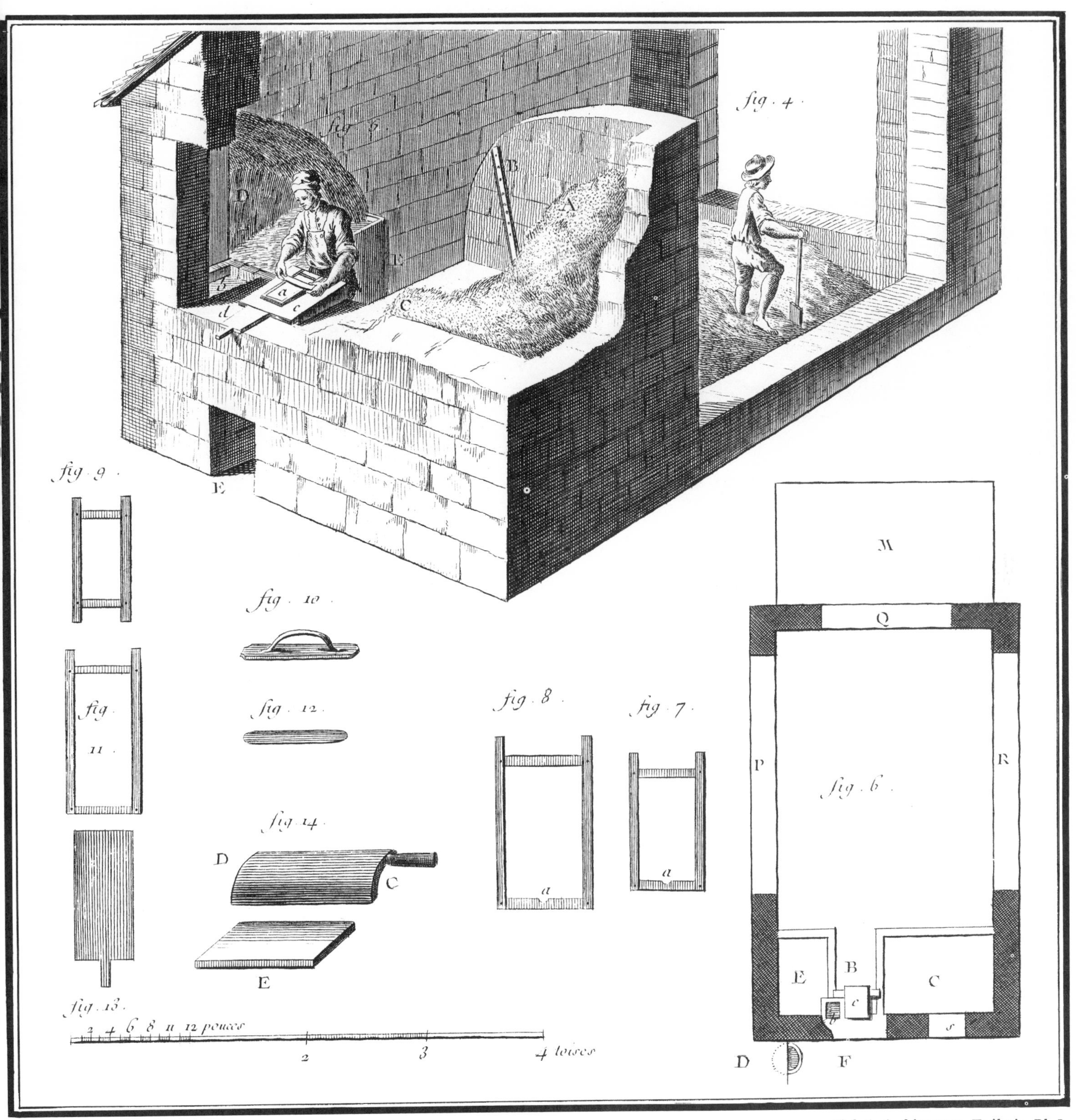
fig. 4.
fig. 9.
fig. 10.
fig. 11.
fig. 12.
fig. 8.
fig. 7.
fig. 6.
fig. 14.
fig. 13.
2 4 6 8 11 12 pouces
2
3
4 toises
M
Q
P
R
E
B
C
D
F

Vol. I, Architecture, Tuilerie, Pl. II.

Plate 279 Tiles II

Drying is completed in the service shed, where unbaked tiles are stacked in lattice-like towers (F) through which the air can circulate. When completely dry, they are cut into hexagonal shape by a pair of tile-cutters, the first of whom halves (Fig. 1) the oblong (A) in which the clay had dried, and hands the squares (C) over to his partner who cuts away the corners. All that now remains is to glaze the tile by firing it in the furnace, from the door of which we are looking into this shop.

Vol. I, Architecture, Carreleur.

Plate 280 Tiles III

Laying tile was a more exact art than the man who walks over it might imagine. It required precise leveling, an exact calculation of the number of hexagonal tiles that would cover a given space, and a precise fitting of one edge to the next. Tiling remains an economical and popular flooring material in modern France.

Plate 281 Tiles IV

Tile competed with slate as roofing in northern France, and where elegance rather than economy governed, tile won out over the more sombre material. But its color was a dull brick red rather than the brilliant orange of Mediterranean roofs.

Here an architect or builder gives instructions to a foreman in charge of a gang repairing a roof. It is probably safe to guess that the contractor, after the manner of his kind, is complaining about the pitch of the roof or some other peculiarly exasperating feature of the job: "If I had known you wanted it that *way . . ."*

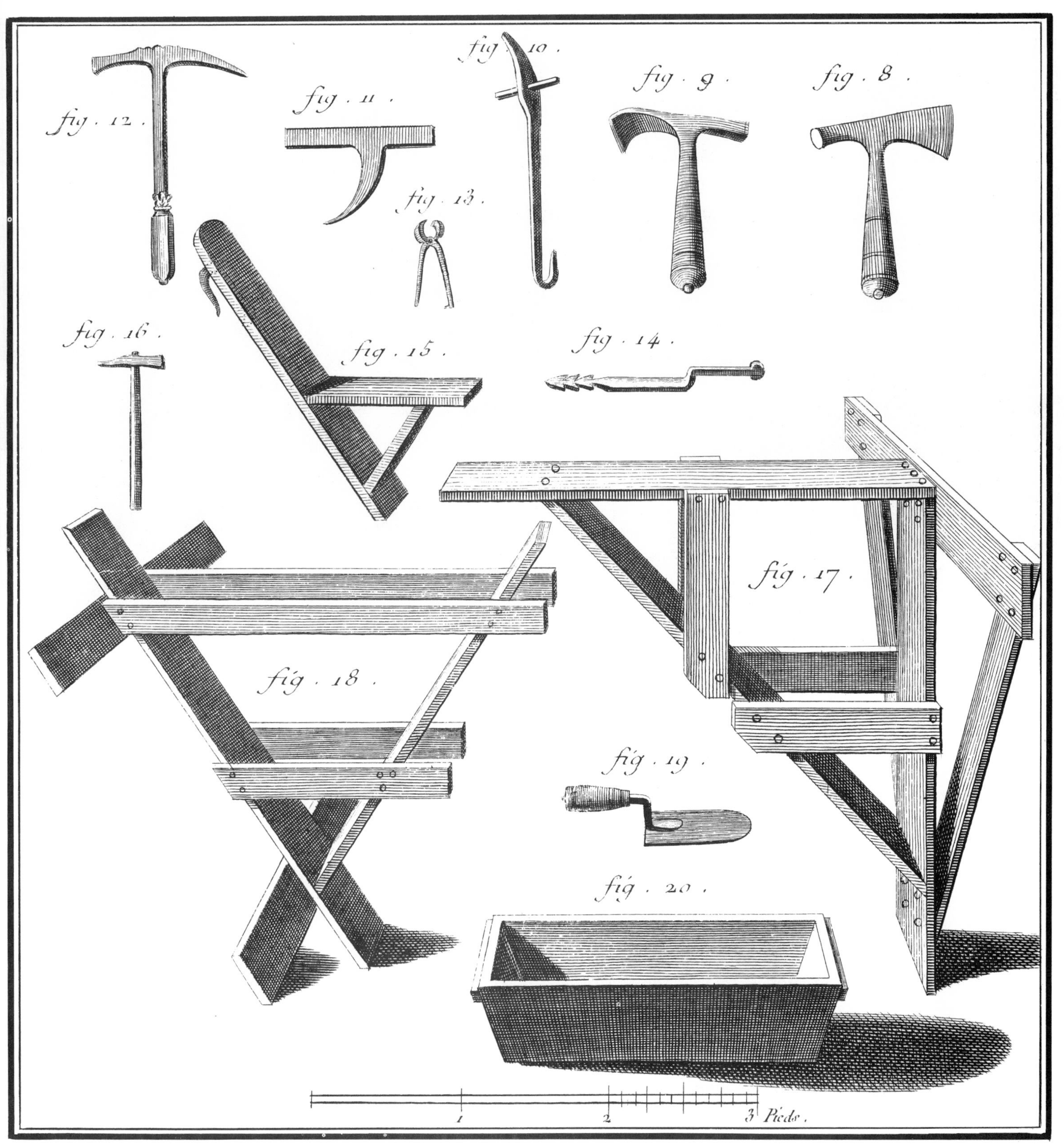
fig. 12.
fig. 11.
fig. 10.
fig. 9.
fig. 8.
fig. 13.
fig. 16.
fig. 15.
fig. 14.
fig. 17.
fig. 18.
fig. 19.
fig. 20.
1
2
3 Pieds.

Vol. V. Marbrerie, Pl. I

Plate 282 Stonework I

The 18th century is often described as an age of restraint when wise men were peculiarly aware of the merit of moderating their enthusiasms. It may have been so, but the modern student is apt to feel that the century did give way to a few enthusiasms with something suspiciously like rapture. One such enthusiasm was for reason. Another, not unrelated though in a different realm (the connection lies in the model of classical civilization) was for marble—the more marble, their architects sometimes seem to have felt, the better. There was scarcely any structure of importance that they would not have faced. It is this marbling touch of the 18th century that is responsible for the extraordinary contents of the choir at Chartres, for example, and not the heavy restoring hand of the 19th century, which is blamed for so many desecrations—often quite rightly.

This vignette is well calculated to illustrate the versatility of marble. Fig. a illustrates a stone saw, with sand dipper. The edge of a stone saw is smooth. Sand and water poured into the groove supply the cutting action.

Plate 283 Stonework II

To choose only one example of the many florid designs suggested for imitation, this is the "improvement" effected by marble in the 13th century floor of the sanctuary and choir of Notre Dame de Paris.

Plate 283

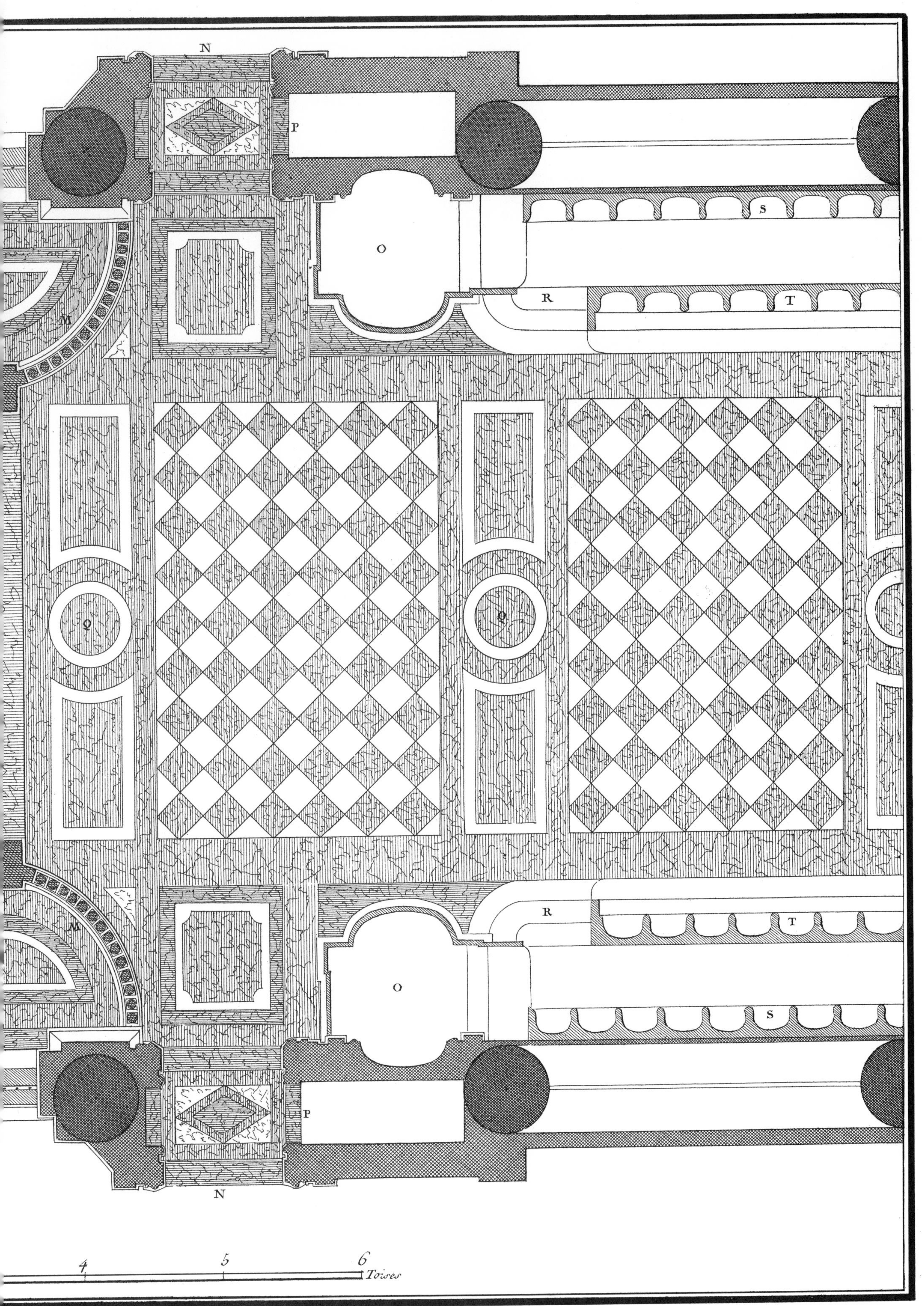
N
P
O
S
R
T
M
Q
Q
R
T
M
O
S
P
N
4
5
6
Toises

Plate 284 Construction

Old styles outlive themselves, and for all their parade of scientific analysis, the Encyclopedists were still close enough to Renaissance habits of thought that they tended sometimes to judge the activities which they treated according to some scale of fancied excellence. They did not really admire carpentry, and regarded it as a regrettable necessity that so much of the framing of buildings should still be of wood. In part their objection was practical: the menace of fire loomed very large in cities unprotected by water systems and heated by open fireplaces and all sorts of dangerous stoves. The smoke of burning London had hung over urban memories since 1666. Beyond this, their disapproval was technological, compounded of belief in rationalization and irritation at the ignorant spirit of routine among carpenters who refused to study the strength of materials or apply mathematical analysis to simplifying, lightening and strengthening their traditional designs.

But however backward, carpentry was a vigorous and fundamental trade, and its tools and methods demanded illustration: (a) sawing timbers, (b) chiseling mortises, (c) squaring joints, (d) trimming down a beam. The workmen are in the charge of a foreman (e) who gets his instructions from the master-builder (notice the difference in dress). Work goes on in a shed (h) in case of rain.

At the left is a timbered front (i) to be plastered. This was a form of building to be deplored, suitable enough for the Middle Ages but not for an advanced culture. The more acceptable use of wood appears at the right (k) in the form of scaffolding for masonry.

Vol. II, Charpente, Pl. I.

Plate 285 Pile Driving I

It was in the workaday world of industry and engineering that carpentry came into its own, for the Encyclopedists could not foresee the steel beam nor even the use of structural cast iron. The following plates will give a few illustrations of what engineers were capable of, given only wood to build with.

This construction has been erected for driving piles into a river bed or harbor to serve as the foundation for a bridge or dock. The hammer (E) has just fallen on a pile which is only started (D). The shelter at the left protects the control mechanism of an underwater saw, the details of which appear in Plate 286.

Plate 286 Pile Driving II

Once the piles are driven into the mud, the design calls for cutting them off under the surface of the water at the level of the sawblade (M) in the left foreground. This may well have been one of the ideal machines designed on paper to stimulate inventiveness rather than to work.

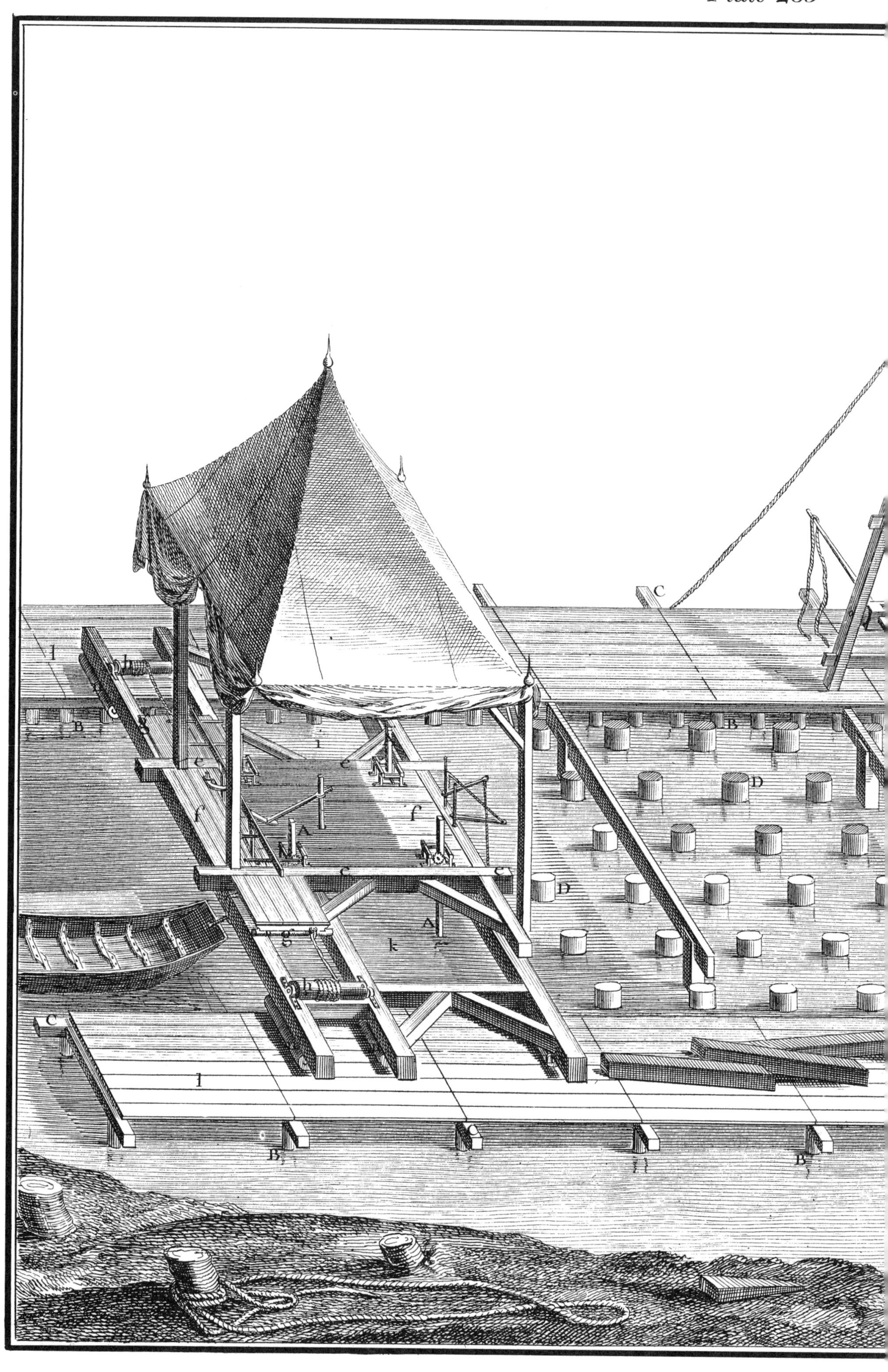
C
l
h
B
B
D
D
e
e
f
f
A
A
c
c
A
g
k
h
C
l
B
B

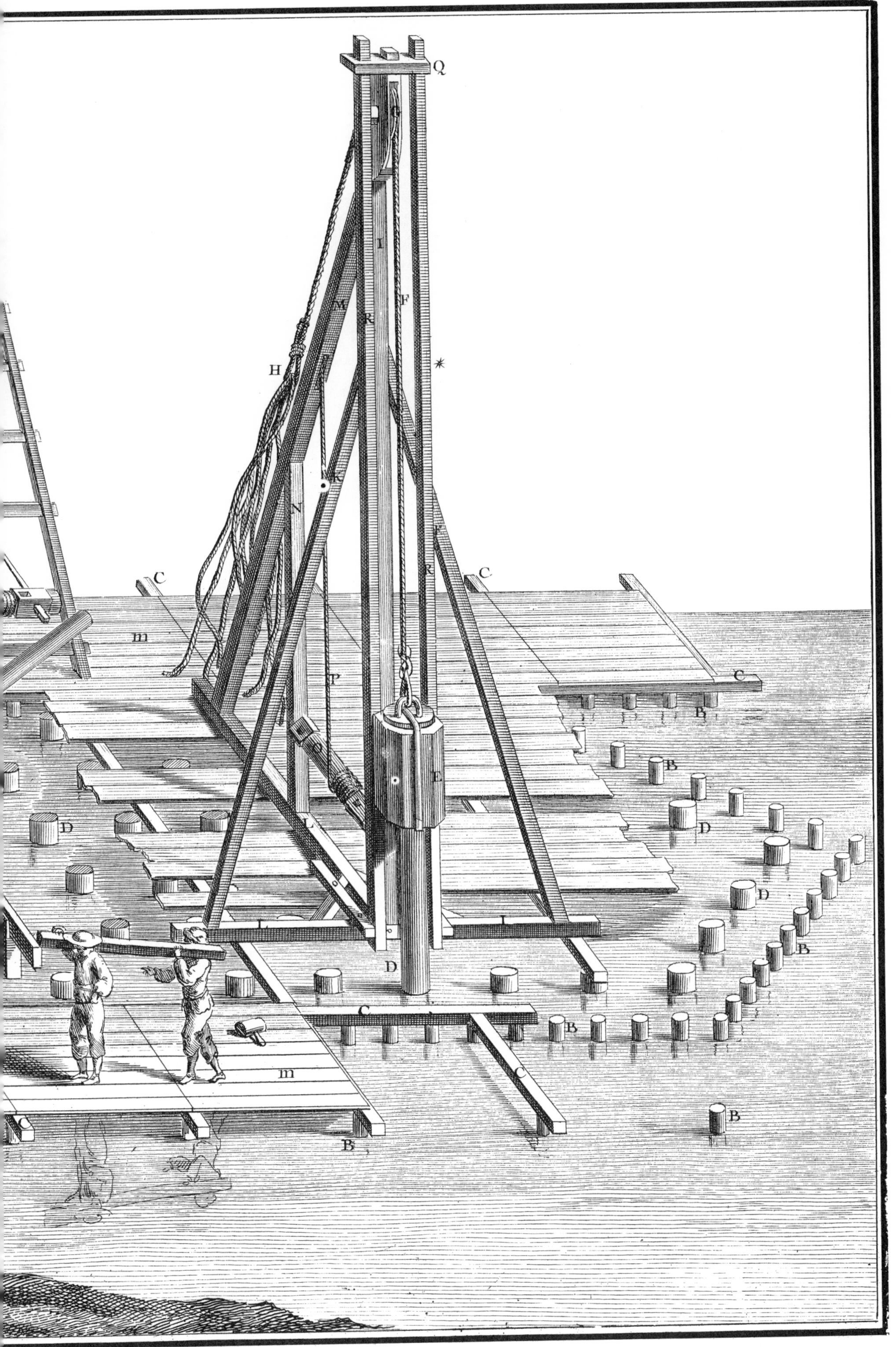

Vol. II, Charpente, Pl. XX.

l
i
c
e
o
b
a
g
f
B
D
k
A
P
d
M
N
K

Pile Driving II

Vol. II, Charpente, Pl. XXI.

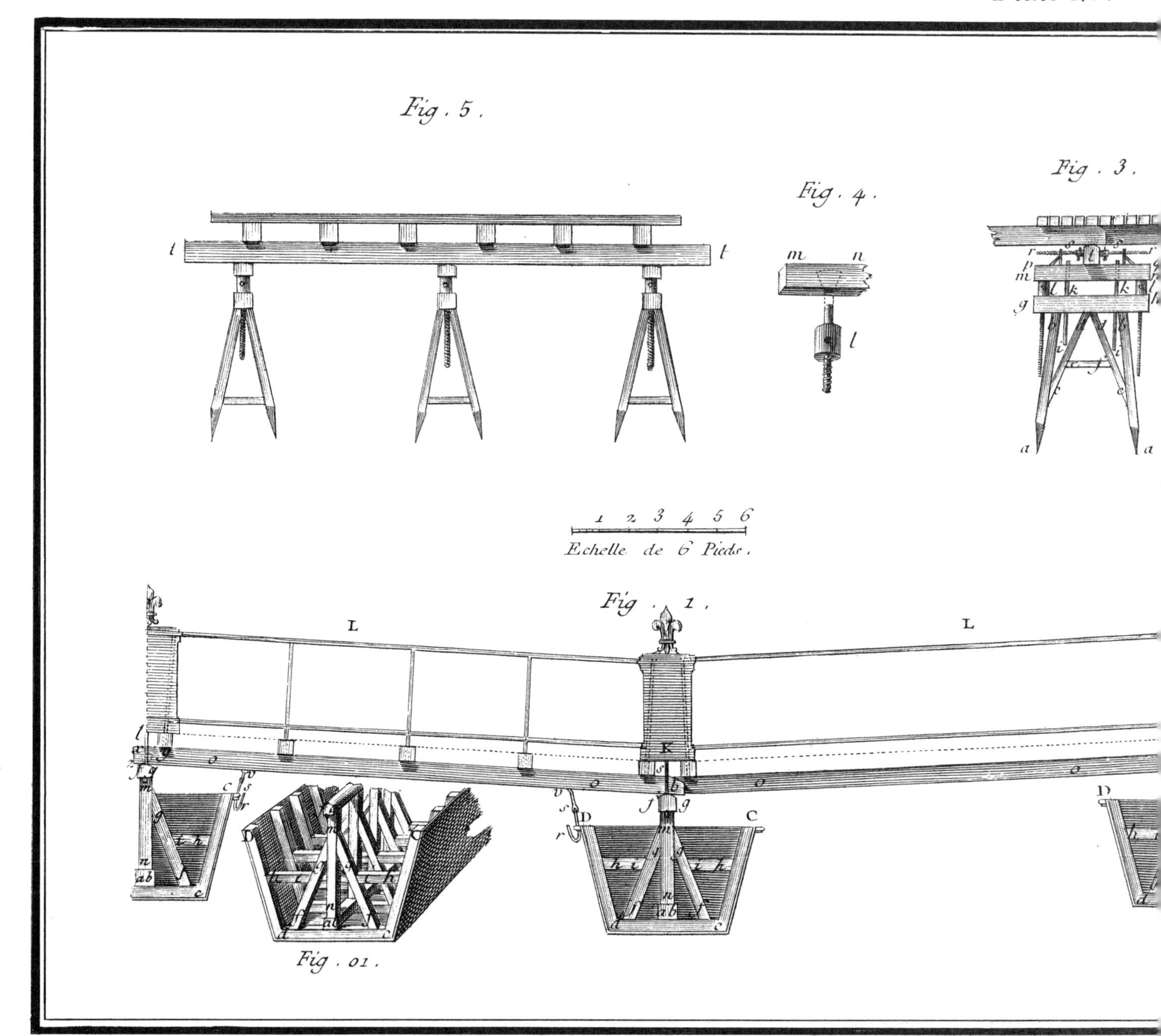

Plate 287 Prefabricated Bridge

A retreating army would normally burn or blow up its bridges, which would have to be replaced by the pursuer. The prefabricated military pontoon bridge of the 18th century military engineer did not differ in principle from that of his World War II successor, although it was perhaps rather more pleasing in design.

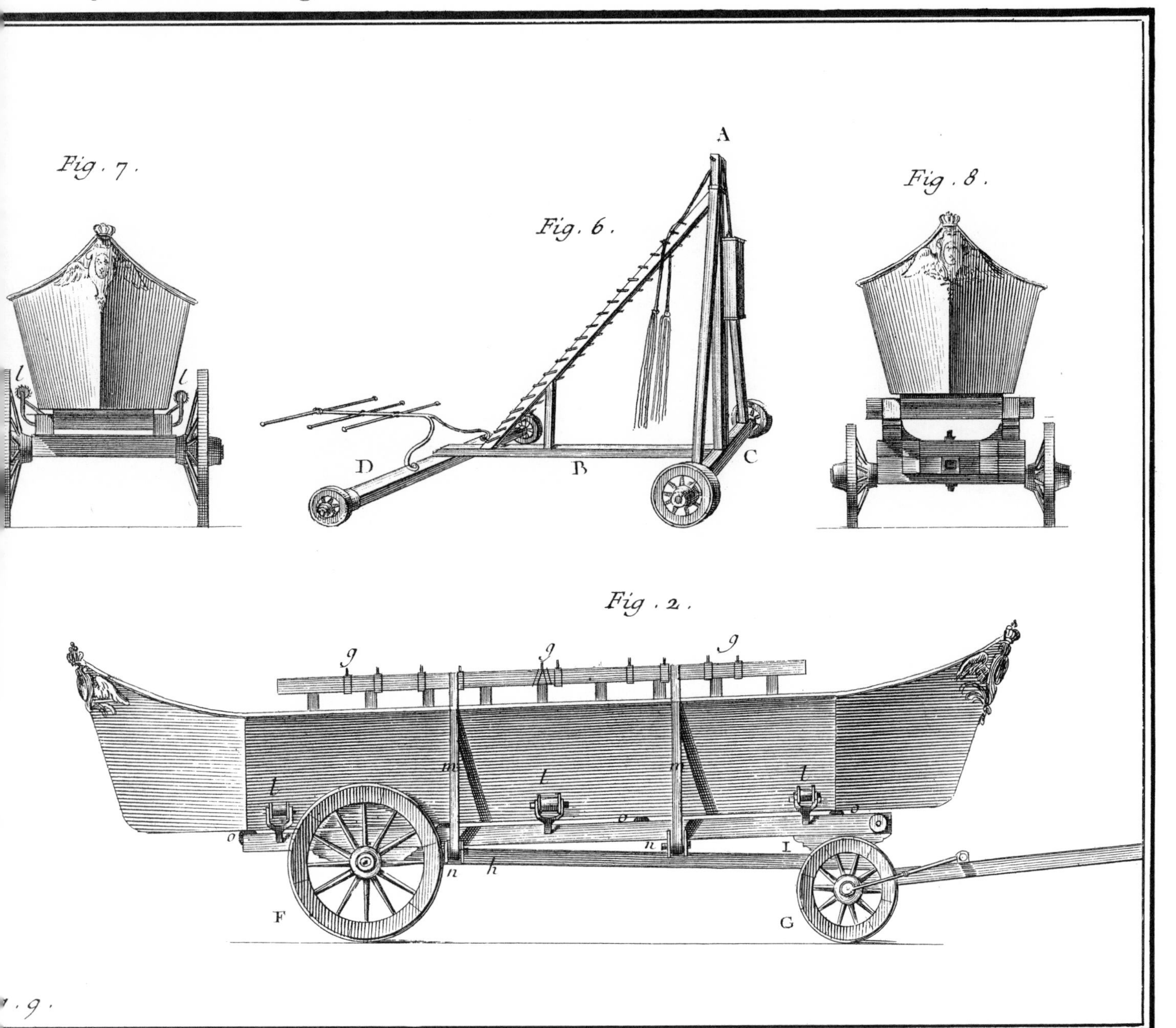
Fig. 7.
l
l
A
Fig. 6.
D
B
C
Fig. 8.
Fig. 2.
g
g
g
m
m
l
l
l
o
o
o
n
n
h
L
F
G
. 9.

D

Prefabricated Bridge

Vol. II, Charpente, Pl. XXIX.

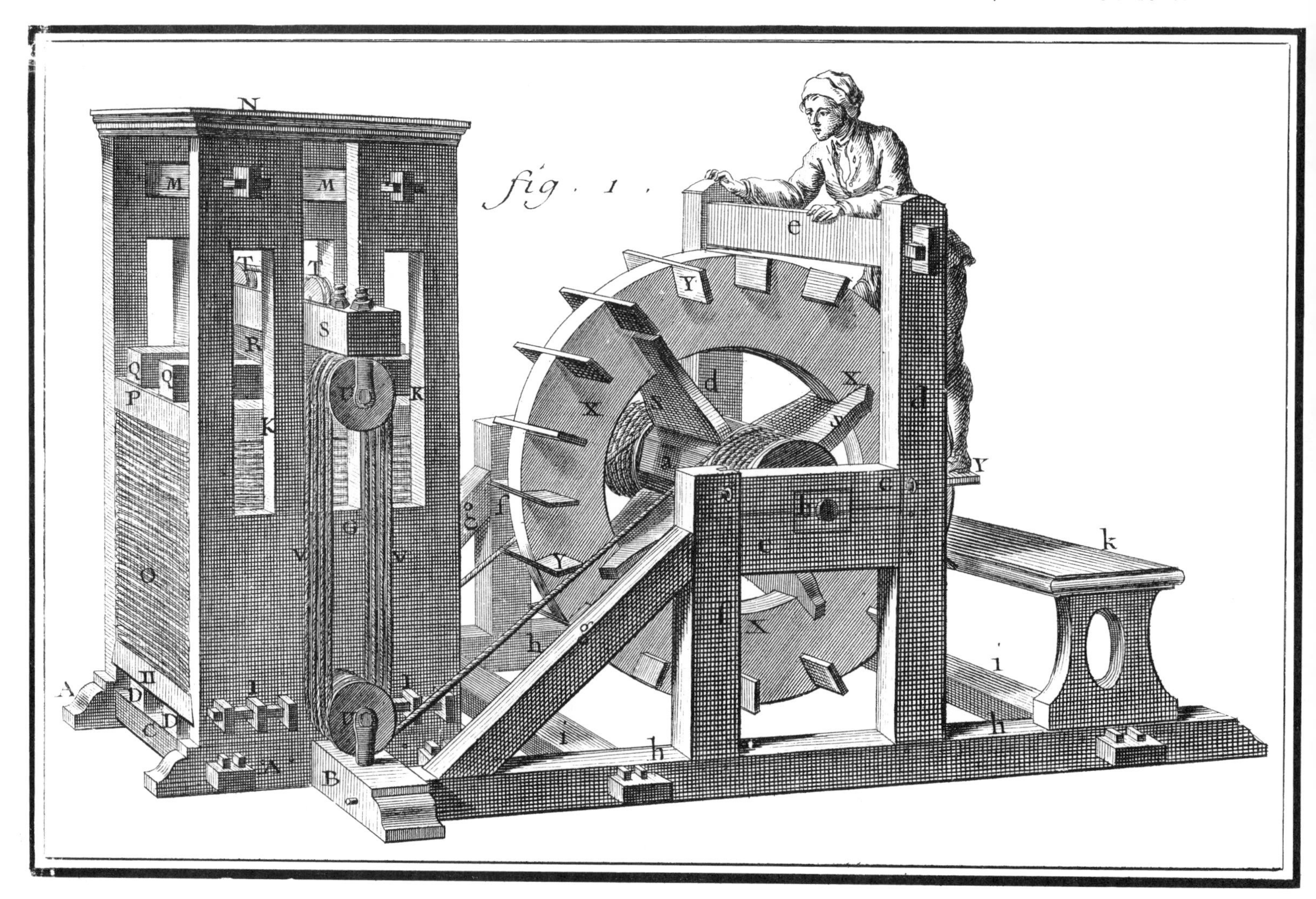

Vol. II, Charpente, Pl. XXXI.

Plate 288 Power I

This vertical treadmill converts the energy which the man expends in climbing endlessly from step to step into considerable power of compression.

Plate 289 Power II

Lifting water by water power itself was the object of a great variety of 18th century pumping devices. The most famous of them, the Machine de Marly is still in operation just west of Paris on the Seine. It was used for pumping water to operate the fountains of Versailles.

The pumping station illustrated here has disappeared. It was situated in Paris itself, at the Pont Notre Dame and in the shadow of the Cathedral. Its purpose is clear enough: to raise water from the Seine to the tanks at the top of the tower (AD).

AD
BB
1
2
3
4
5
6 Toises
O
X
R
R
N
N
I
I
P
P
Q
Q
C
C
G
G
Y
Y
X
X
Y
M
M
K
K
F
F
E
E
L
L

Plate 290 Shipbuilding

One tends to forget, sometimes, that the French have a maritime tradition second only to that of England, and that there have been times when English admirals have had reason to complain of the quality of French ships. Shipbuilding required all the resources of which the art of carpentry was capable.

The plate shows the latest model of drydock. One is to imagine the keel of the ship under construction as laid on the joists (FE) notched for it. The artist has moved it up onto the shore only to let us see the arrangement.

There remain certain mysteries about this plate, however—for example, the sages in ancient garb, two of whom stand disputing on the far side. A third is at the head of the steps discussing the drawing of a boat of obviously obsolete design. Who are they—certainly not 18th century naval architects—and what do they symbolize?

B
A
B
L
K
N
C
D
E
F
G
H
H
H
H
N

M
L
K
L
K
N
N
H
F
E
G
H
N

Vol. IV, Ebénisterie et marqueterie, P

Plate 291 The Cabinetmaker

Since the 18th century the tools and methods of the carpenter-builder have evolved in just that direction of lightness and precision toward which they were urged by the Encyclopedists. The cabinetmaker's techniques have changed much less. Indeed, when department store furniture is compared to what could be bought by customers of comparable establishments in the 18th century, it is clear that the change involves the deterioration of a trade, not its development.

The plate represents a cabinetmaker's workroom, strewn with tools and materials in that state of disorder which sometimes bespeaks confident craftsmanship.

Vol. VII, Menuiserie, Pl. I.

Plate 292 The Joiner I

By modern standards division of labor and specialization of function were still at a fairly elementary level in the 18th century. Specialization of trades, on the other hand, was carried farther than it is now and there were many more different kinds of carpenters. Carpentry proper meant gross constructions. For interior work, one would hire not a carpenter, but a joiner, whose specialty was, as the name implies, fitting different pieces and kinds of wood for window frames or door frames, parquet floors or panelled walls. This is a joiner's yard. Two of his men are using a ripsaw.

Vol. VII, Menuiserie, Pl. II

Plate 293 The Joiner II

Inside the shop, one sawyer handles the ripsaw (a) and another the crosscut saw (b). At the far worktables the two-man plane (e) and brace-and-bit (d) are in use. At the left (Fig. f) a piece of parquet flooring is being finished, and completed articles of joinery stand about awaiting delivery (g, h).

Plate 294 Chairs

Vol. VII, Menuisier en Meubles, Pl. I.

Plate 294 Chairs

The furniture industry was divided among a number of trades. The cabinetmaker constructed chests, armoires, cupboards—but making chairs was another of the jobs belonging to joiners, all pounding away on top of each other. The boy with the brazier is melting glue for the joints. No doubt the congestion of this scene owes something to the artist's desire to show in one picture everything that joiners make.

Vol. VII, Menuisier en Voitures, Pl. I.

Plate 295 Carriages I

The bodywork on carriages pertained to joiners specializing in vehicles. At the moment they are making a berlin.

Plate 296 Carriages II

Vol. IX, Sellier-Carossier, Pl. I.

Plate 296 Carriages II

Besides joiners, the carriage-maker also employed harness-makers and upholsterers. He himself would vary the design to suit his customers' taste and status, for the standard body types of the 18th century carriage permitted the owner to strike almost any note from utilitarian sobriety to sumptuous display or gay frivolity.

Plate 297 Carriages III

The berlin seats its passengers face to face. This berlin is a lady's town carriage.

Plate 298 Carriages IV

A country berlin.

Plate 299 Carriages V

A calèche.

Plate 300 Carriages VI

The diligence de Lyon *is all business, like the city from which it takes its name.*

Plate 301 Carriages VII

The Diable *combined speed with style.*

Plate 302 Carriages VIII

The post-chaise was a two-wheeler.

Plate 303 Carriages IX

The cabriolet was also a two-wheeler. It is apparent why this name has been appropriated by manufacturers of sporty cars.

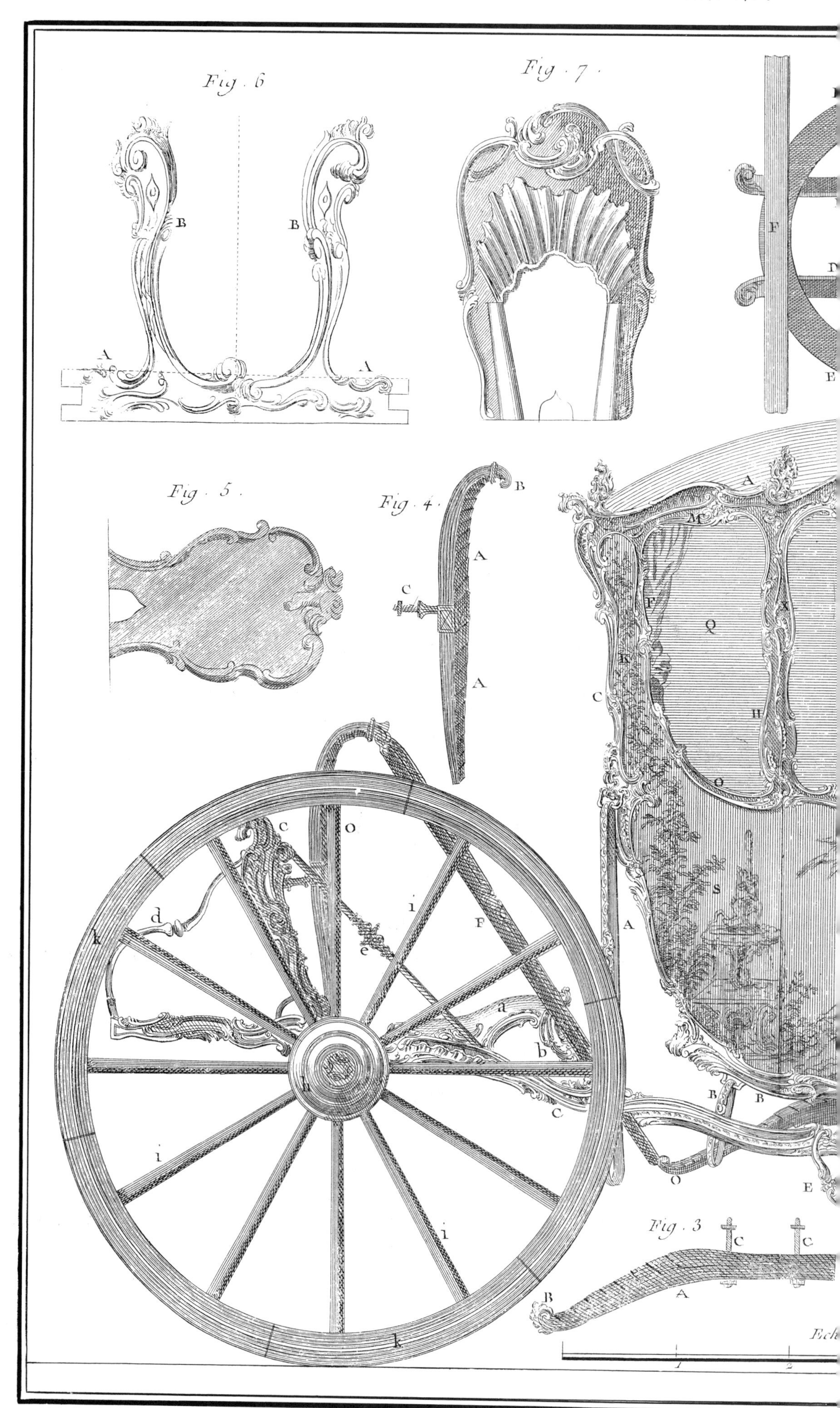
Fig. 6
B
B
A
A
Fig. 7.
F
D
E
Fig. 5.
Fig. 4.
B
A
C
A
A
M
F
X
Q
K
C
H
O
S
A
c
o
d
i
k
F
e
a
b
h
C
B
B
O
E
i
i
k
Fig. 3
C
C
B
A
1
2

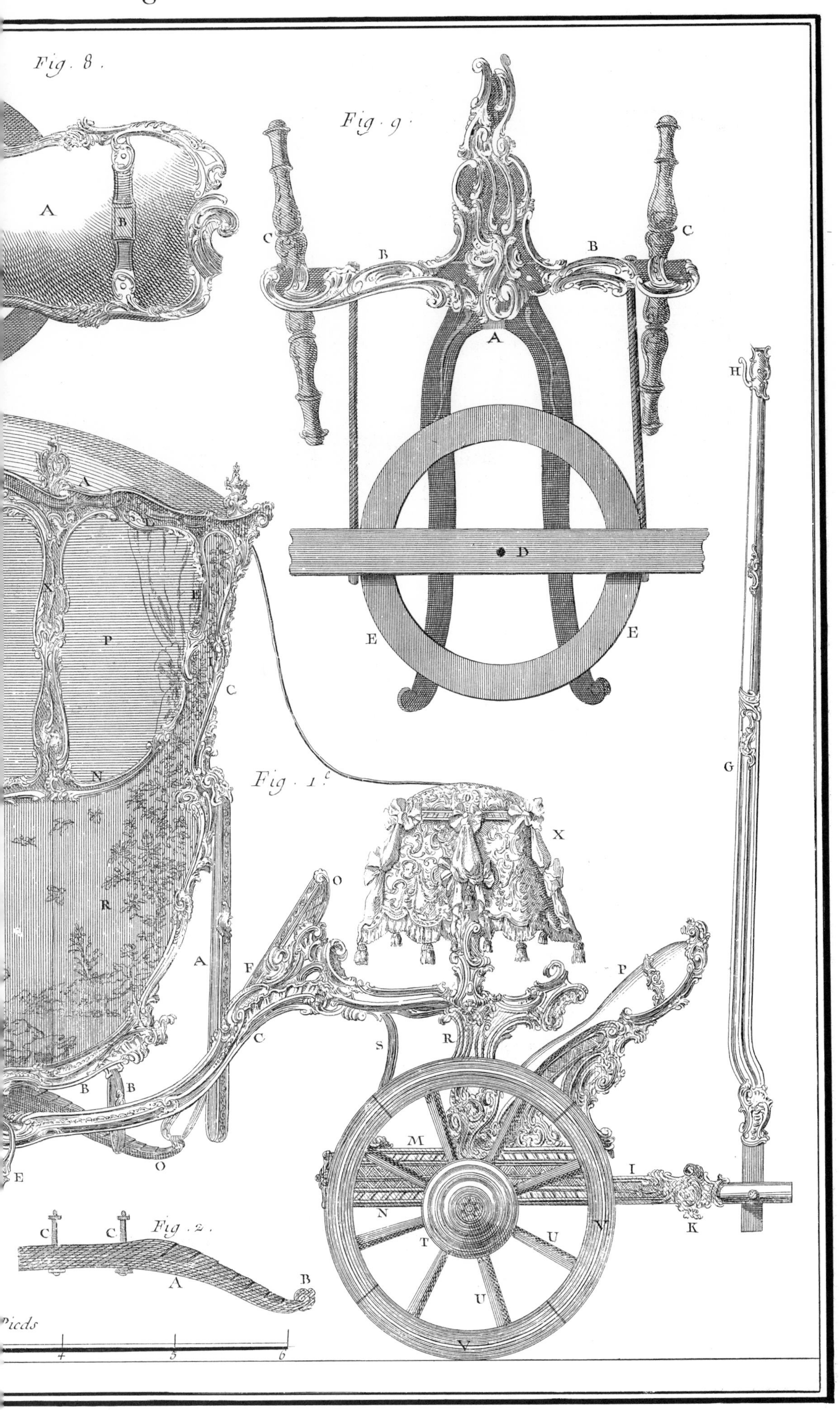

Fig. 8.
A
B
Fig. 9.
C
B
B
C
A
H
D
E
E
X
P
C
N
Fig. 1.
G
R
O
A
F
P
C
S
R
B
B
M
O
E
I
N
V
K
T
U
C
C
Fig. 2.
A
B
U
Pieds
4
5
6
V

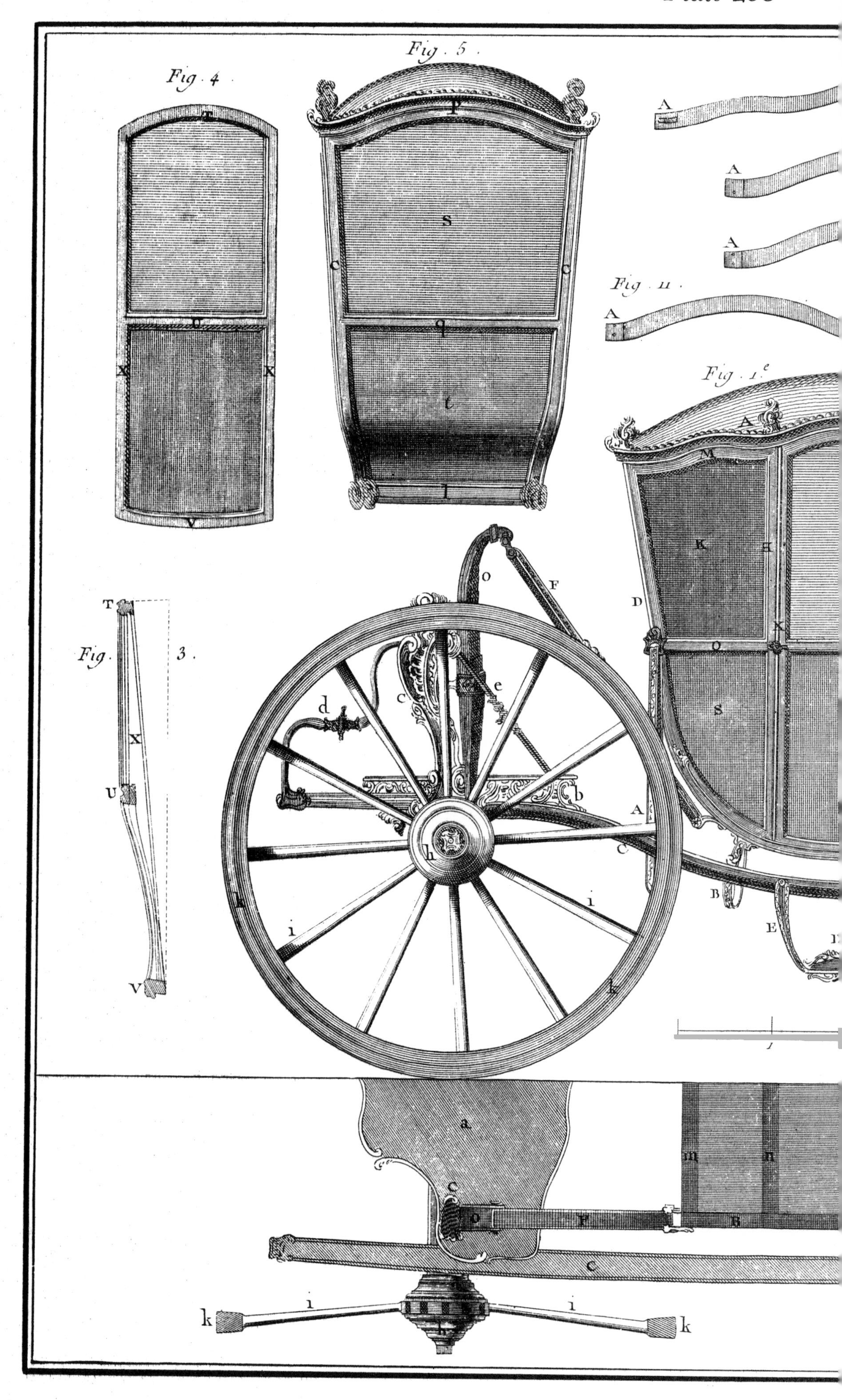

Fig. 5.
Fig. 4.
Fig. 11.
Fig. 1.e
Fig.
3.

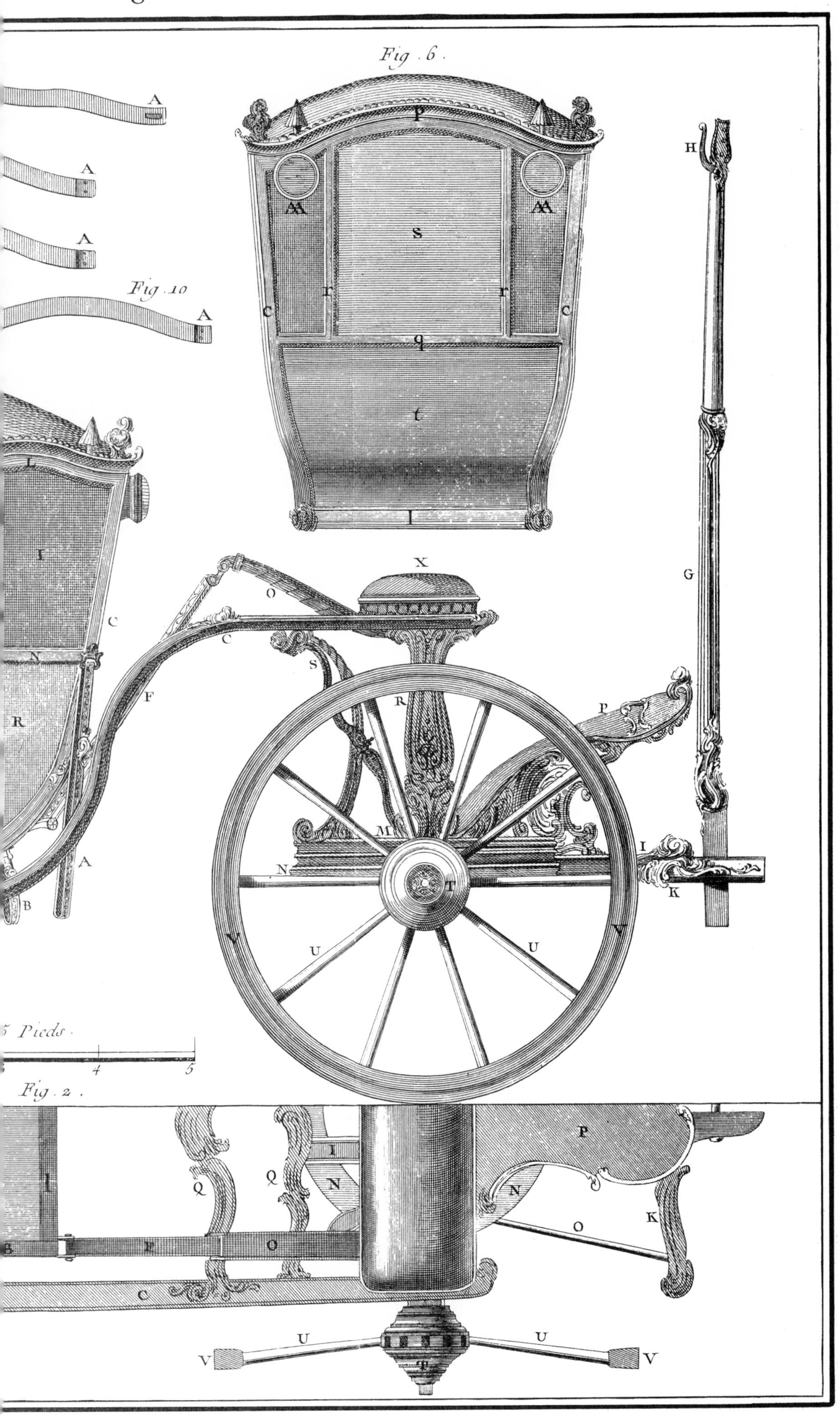
Fig. 6.
Fig. 10
Fig. 2.
5 Pieds.
4
5

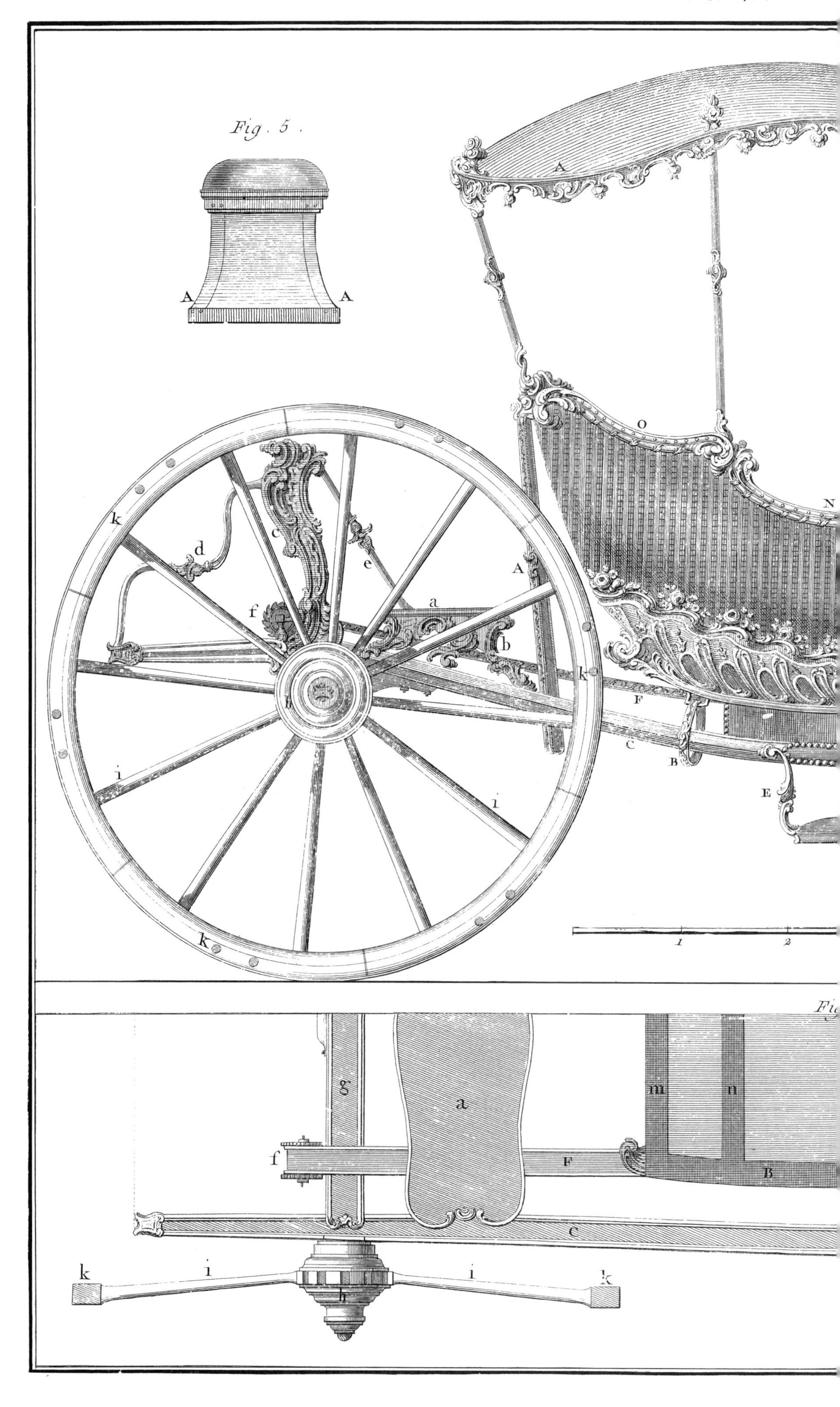
Fig. 5.
A
A
A
O
N
k
d
c
e
A
a
f
b
k
F
h
C
B
E
i
i
k
1
2
g
a
m
n
f
F
B
C
k
i
i
k
h

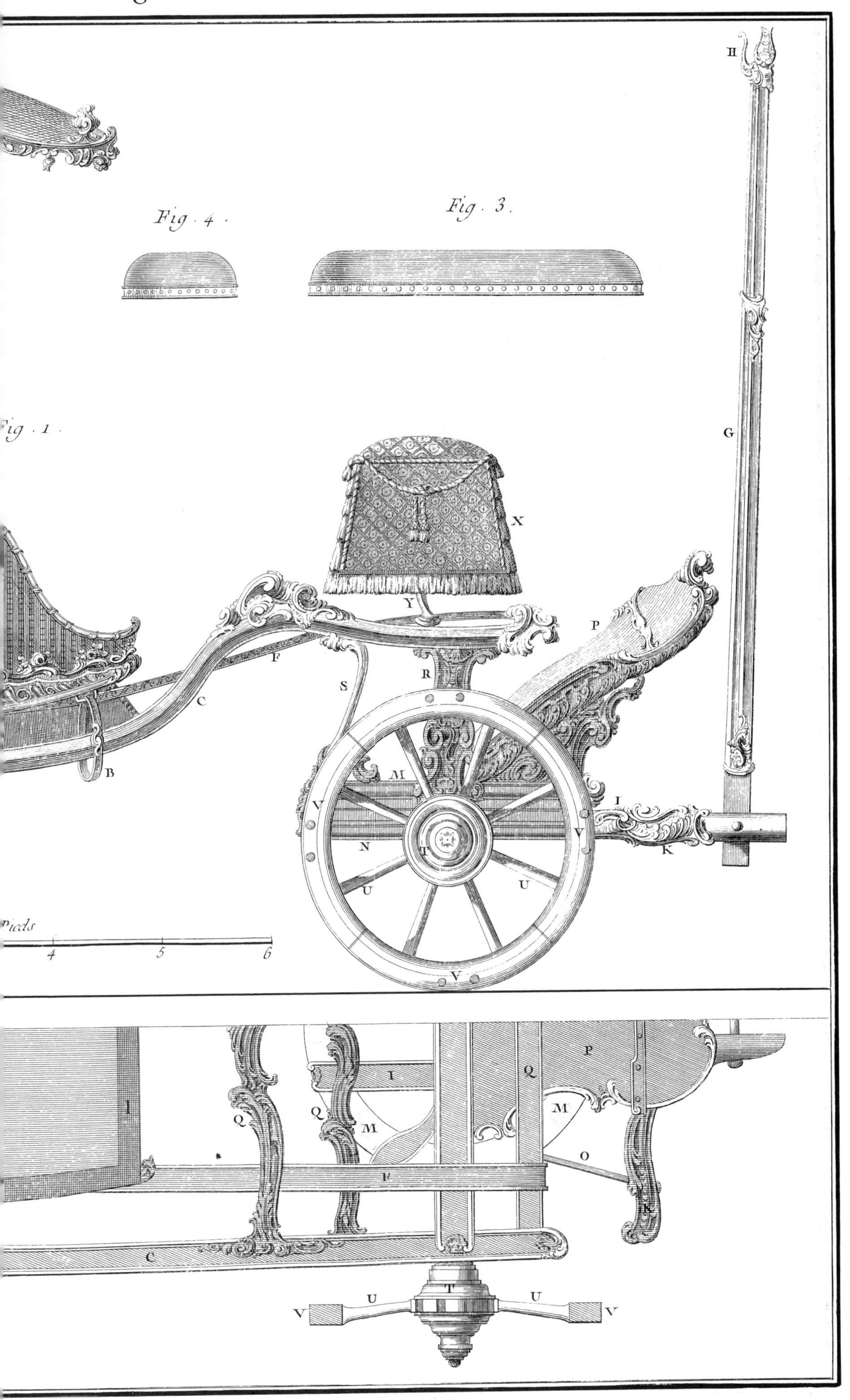

Fig. 4.
Fig. 3.
Fig. 1.
H
G
X
Y
P
F
R
S
C
B
M
I
V
N
T
K
U
Pieds
4
5
6
Q
O

Fig. 3.
Fig. 6.
p
s
r
c
q
t
A
M
F
K
H
G
S
D
X
d
i
k
h
B
O

Fig. 7.
Fig. 5.
A
A
H
E
F
D
A
B
C
E
F
G
y
X
P
I
F
G
N
C
A
X
R
B
u
V
O
C
I
A
T
U
U
V
V
E
C
Q
Q
I
I
F
K
U
U
V
V
T
Pieds.
4
5
6
Pieds.

Fig. 4.
A
A
A
i
i
O
A
D
S
f
a
F
k
i
h
i
B
C
k
a
m
n
g
g
f
F
C
i
i
k
h
k

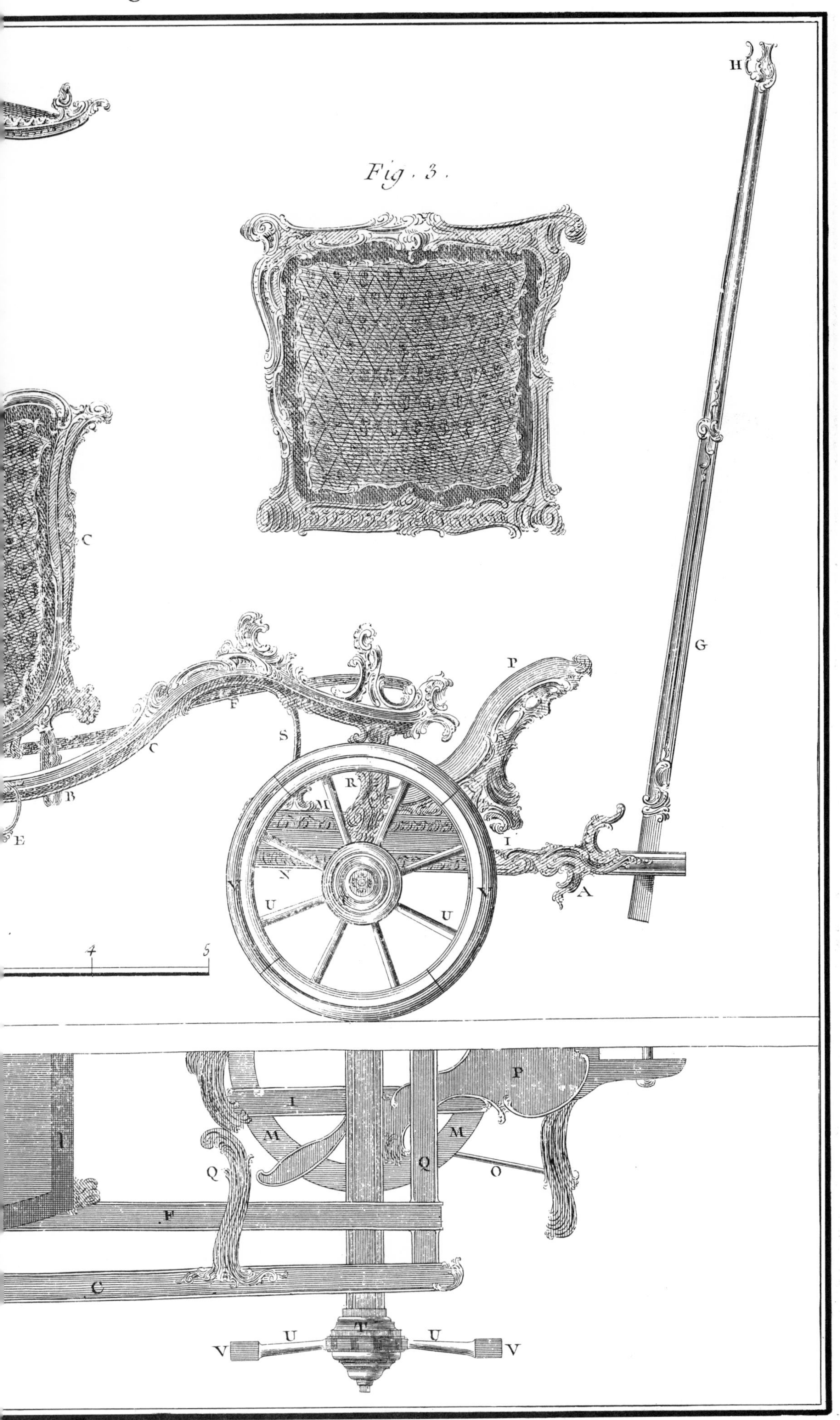
Fig. 3.
H
G
C
P
F
S
C
B
E
R
M
I
N
V
U
A
4
5
P
I
M
M
Q
Q
O
F
C
T
U
U
V
V

Fig. 12.
A
A
C
B
C
Fig. 11.
A
A
A
A
Fig. 1e
Fig. 3
A
A

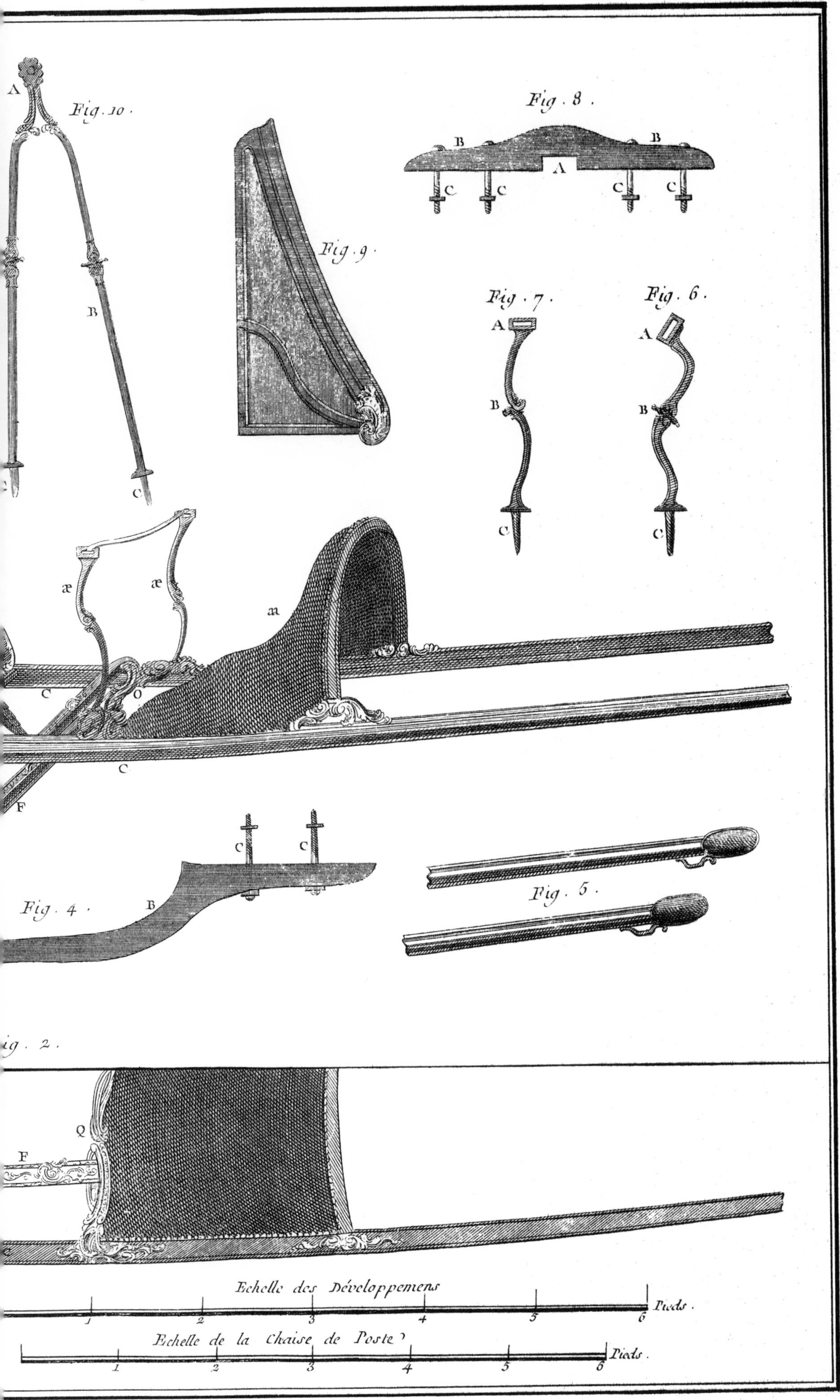
Fig. 10.
A
B
C
Fig. 9.
Fig. 8.
B
B
A
C
C
C
C
Fig. 7.
Fig. 6.
A
A
B
B
C
C
æ
æ
aa
C
O
C
F
C
C
Fig. 4.
B
Fig. 5.
ig. 2.
Q
F
Echelle des Développemens
1
2
3
4
5
6
Pieds.
Echelle de la Chaise de Poste
1
2
3
4
5
6
Pieds.

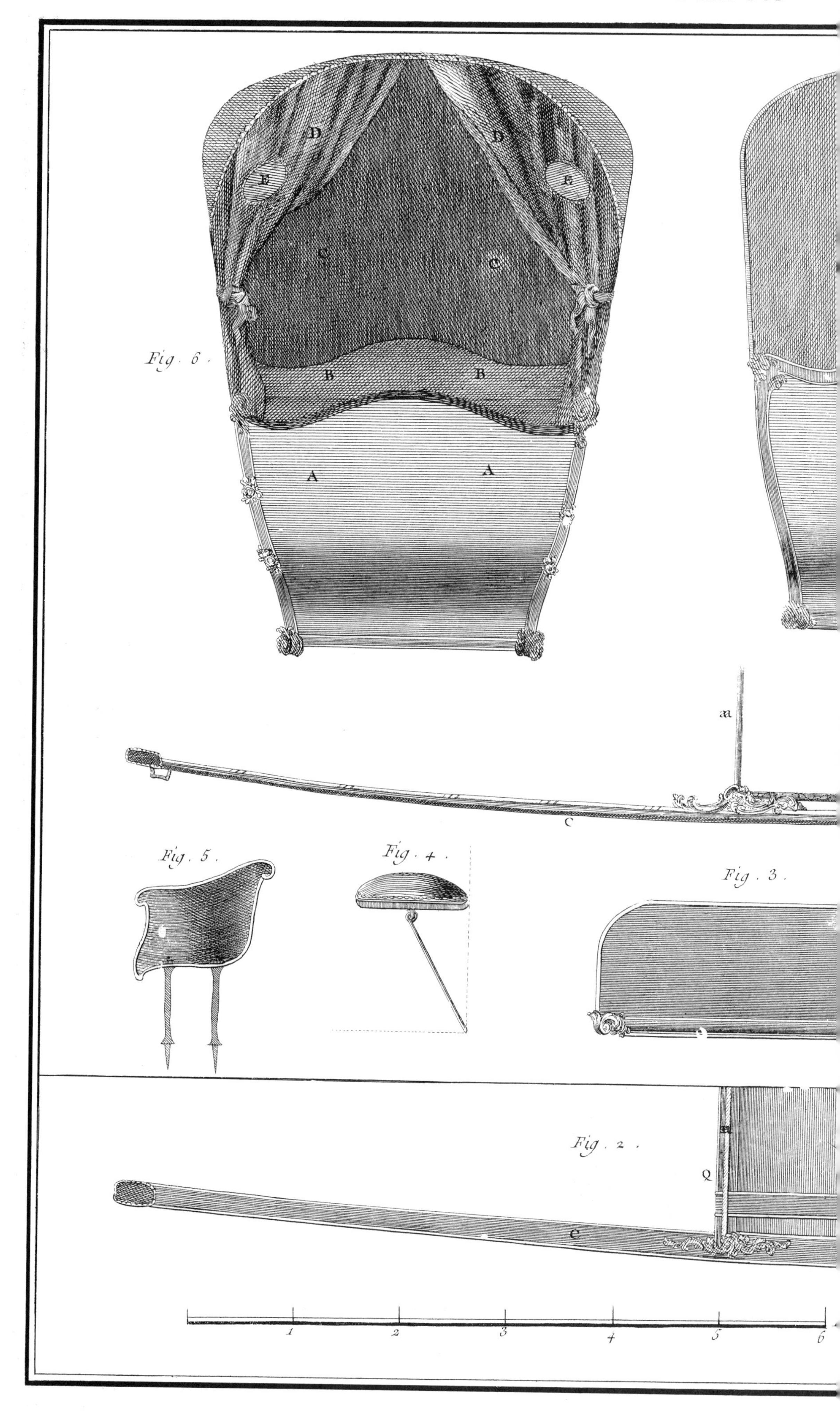
Fig. 6.
D
D
E
E
C
C
B
B
A
A
æ
C
Fig. 5.
Fig. 4.
Fig. 3.
Fig. 2.
Q
C
1
2
3
4
5
6

Fig. 8.
Fig. 7.
Fig. 9
Fig. 1e.
A
B
C
D
E
F
G
H
I
K
L
S
a
b
f
h
i
k
g
L
m
n

Textiles

Plate 304 Wool I

Later history showed that the textile trades were the ripest of all industries for technical revolution. This circumstance renders the Encyclopedia's *sections on spinning and weaving somewhat ironical reading, for eager as the authors were to exemplify progress, the effect is more archaic than elsewhere, and they exhibit not the methods of the future, but the final evolution of techniques about to be submerged in the chain reaction that led from flying shuttle through spinning jenny, water-frame, and mule to power loom.*

These developments originated in the English cotton industry, from where they spread by adaptation to other textiles and other countries. The first plates portray French woolen workers. This was not one of France's strongest trades. Her flocks were never the equal of the famous merinos of Spain, nor did her clothiers threaten the reputation of Yorkshire.

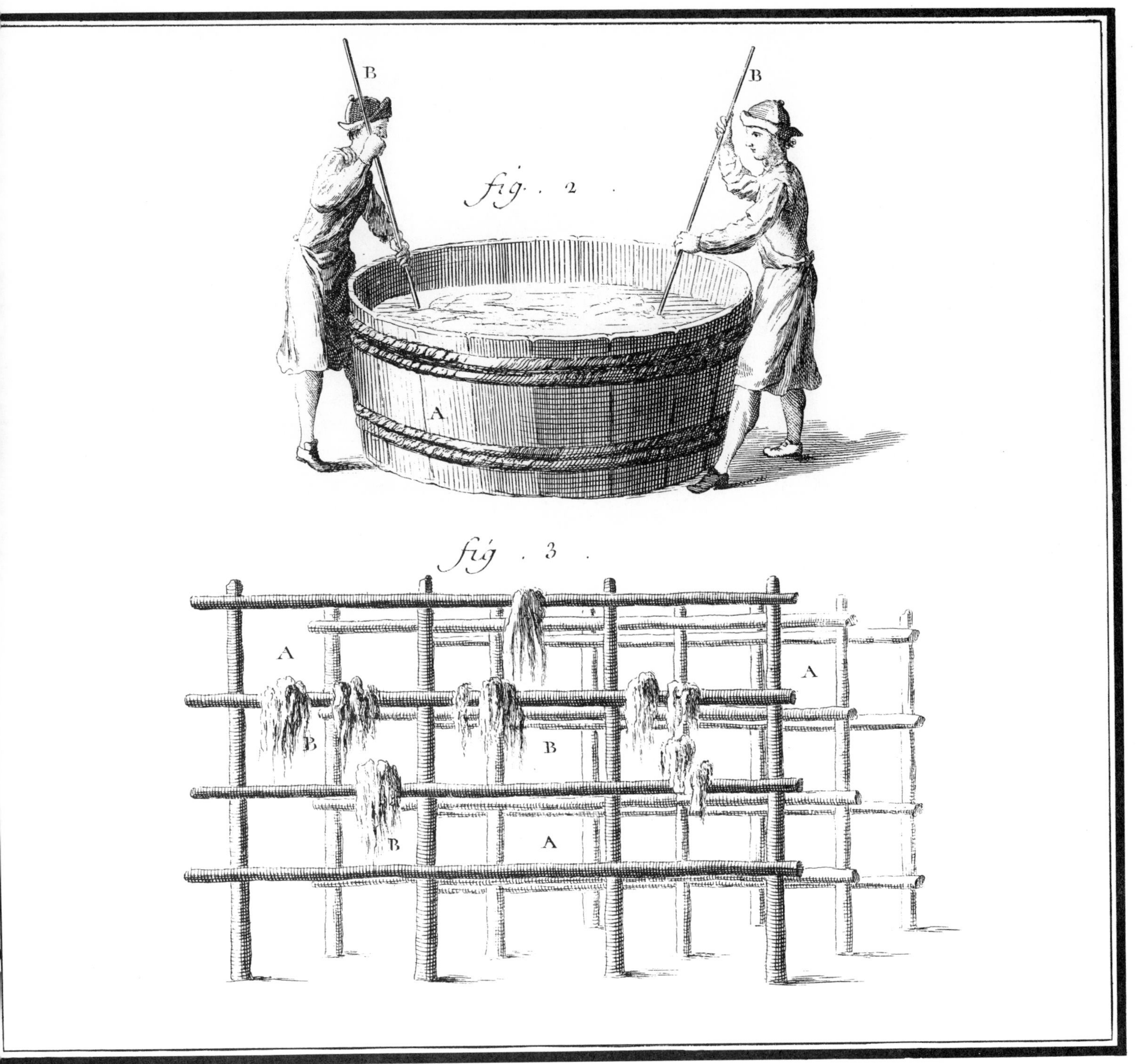

Vol. III, Draperie, Pl. I.

Indeed, it is only the first of these plates, in which the work seems the most primitive, that has an air of authenticity. In the background is a shearing barn, before which laborers are washing bundles of fleece in a tub containing three parts tepid water and one part urine. This rids it of grease and encrusted salts. The workers rinse the wool in running water, and dry it in small bunches. That destined for flannel is given a second washing (Fig. 2) in soapy water.

Plate 305 Wool II

Before being spun, wool is picked (Fig. 4) and the fibres sorted according to length and quality. They are then beaten to rid them of dust (Fig. 5).

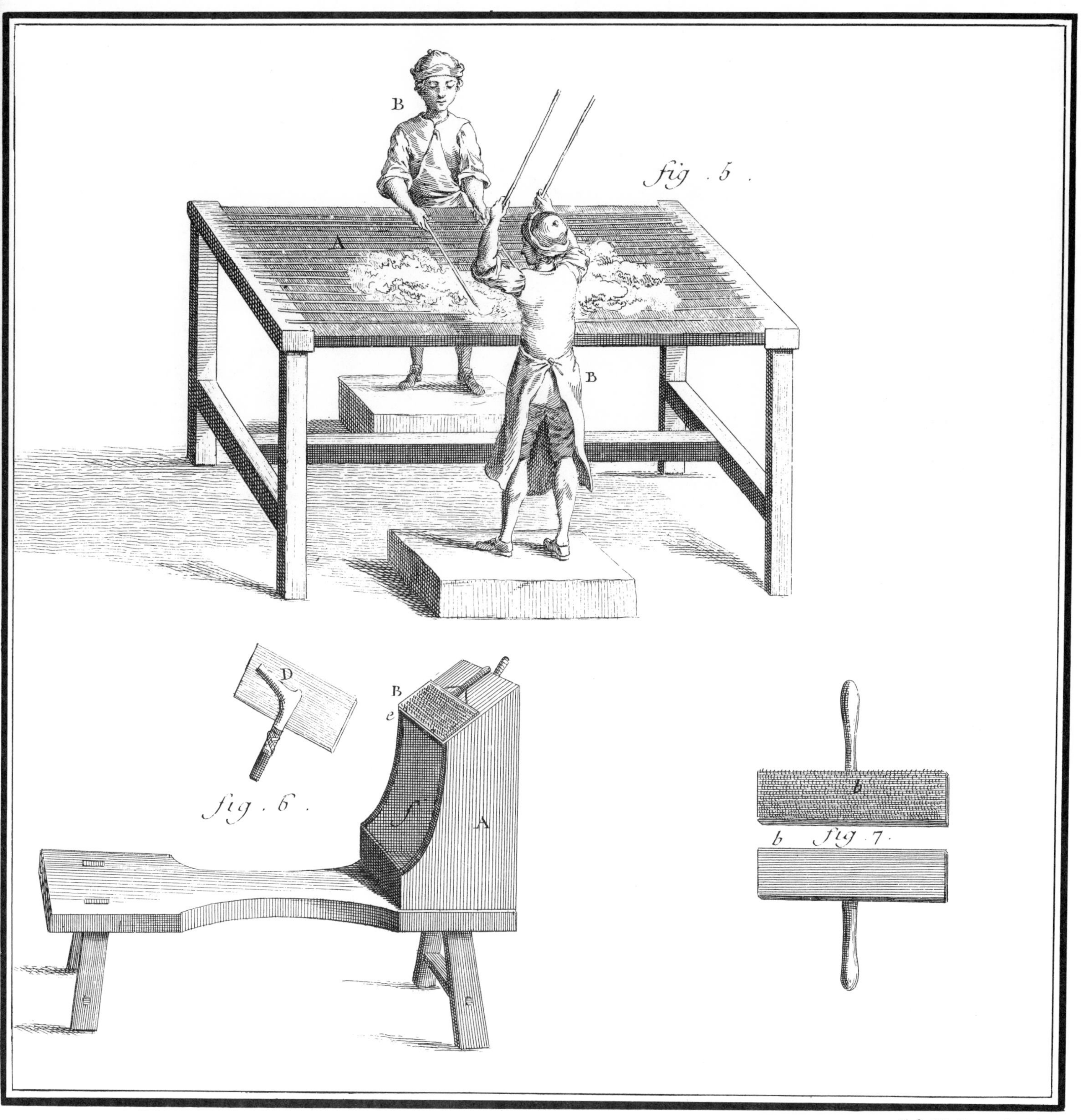
B
fig . 5 .
A
B
D
B
e
fig . 6 .
f
A
b
b fig 7.

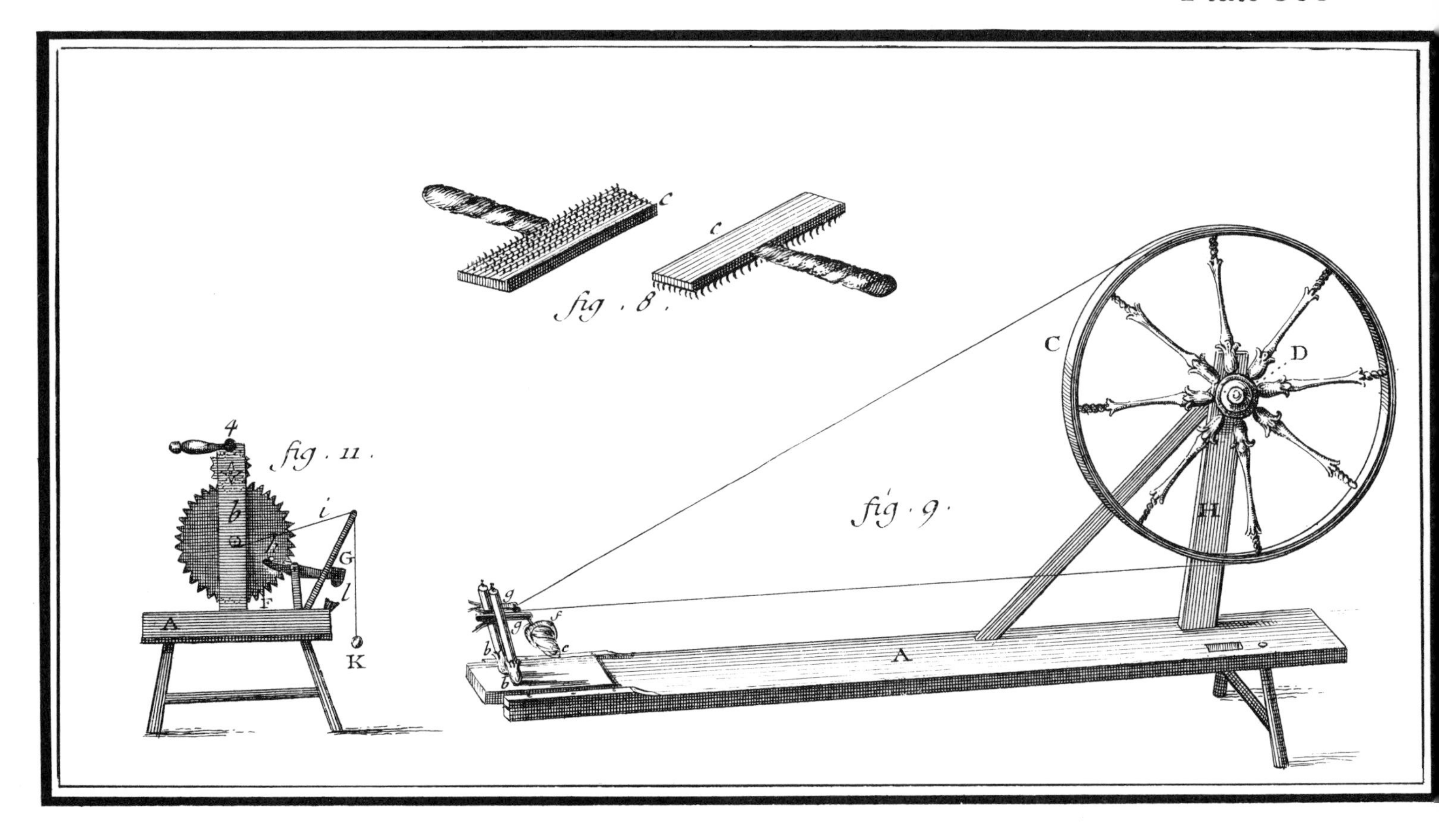

Plate 306 Wool III

The wool-carder (Fig. 8), spinning wheel (Fig. 9) and reel (Figs. 10, 11) are familiar implements. In Fig. 12 the spinner is sizing wool. He runs it through a solution made by boiling rabbit skins or old gloves to a pasty, almost gelatinous consistency. Then he spreads the loosely twisted hanks on racks to dry.

fig. 10.
fig. 9. No. 2.
fig. 12.
fig. 13.

Plate 307 Spinning I

This spinning scene is rather more lifelike, and more ladylike. Spinning remained an occupation of the housewife long after weaving had become a trade. No doubt this was because a loom required a greater investment.

These good bourgeois wives have brought together the main domestic spinning devices. The apprehensive girl at the left (Fig. 1) is spinning clumsily, with spindle and distaff, different types of which are illustrated opposite, Figs. 5-7. At the right is the more familiar spinning wheel, operated by a treadle. It was more expensive as well as more productive. An enlargement appears in Fig. 10. The woman in the center (Fig. 3) winds a skein, and her companion (Fig. 4) a ball of yarn.

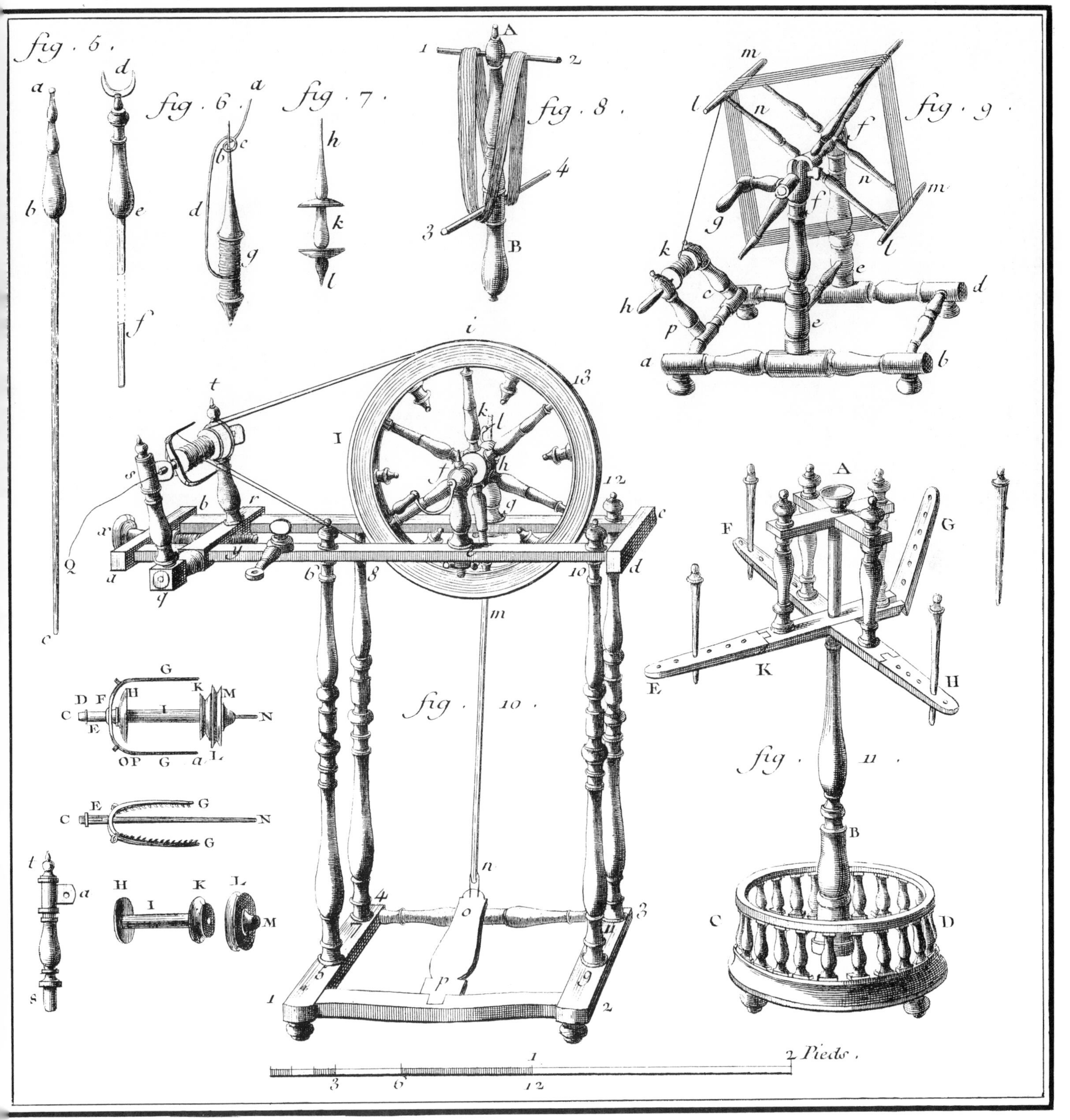
fig. 5.
fig. 6.
fig. 7.
fig. 8.
fig. 9.
fig. 10.
fig. 11.
2 Pieds.

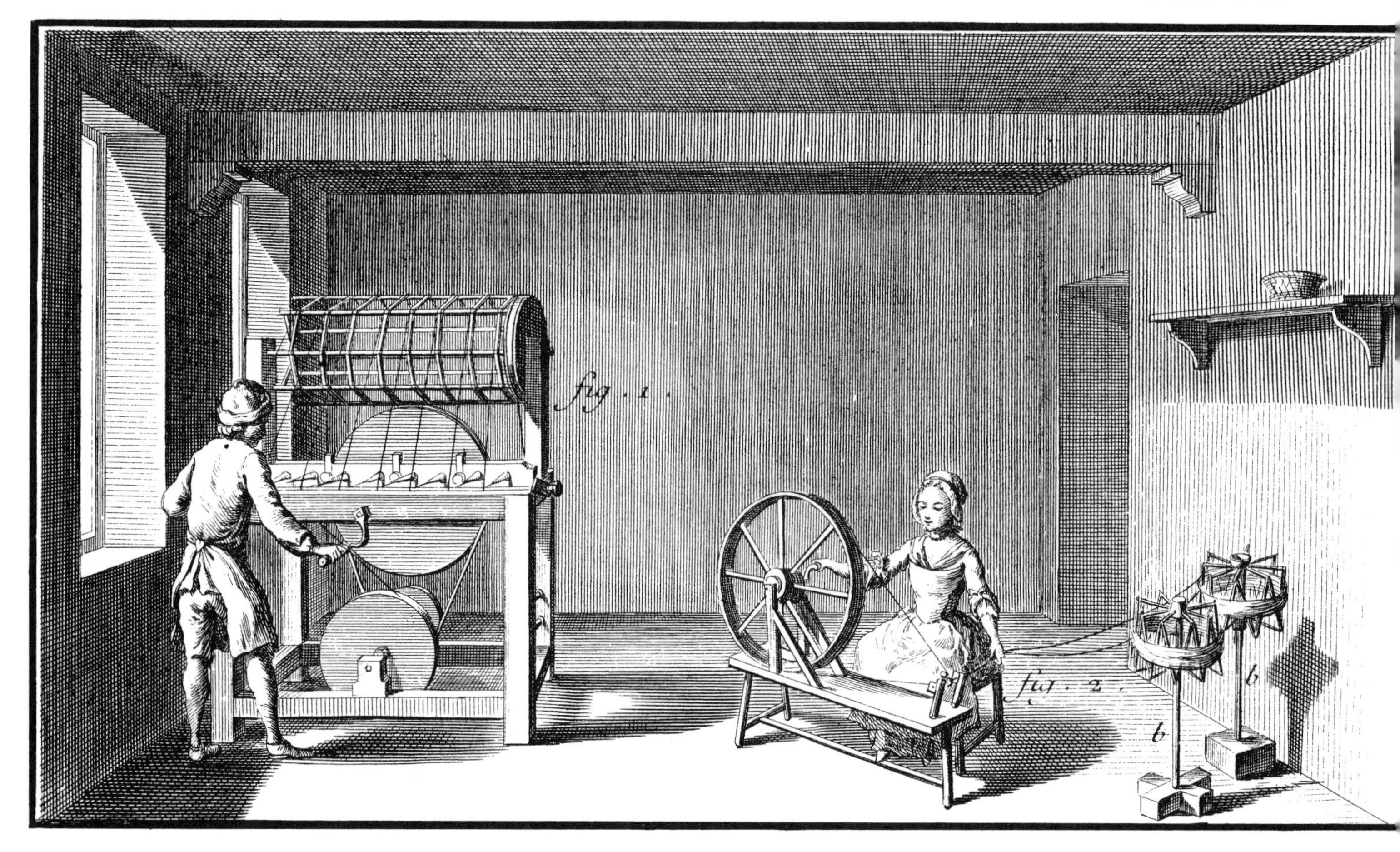

Plate 308 Spinning II

These spinners have the air of employees once again. The purpose of the device at the left (Fig. 1) is left in some obscurity by the Encyclopedia, *although its actual operation is clear enough from the detailed illustrations opposite (Figs. 3, 4). Yarn is wound onto the reel from bobbins which turn at right angles to it. This imparts added twist to the threads. Tension is maintained by virtue of the differing speeds at which reel and bobbins revolve. It is probable that this device is a warping reel. In some establishments the warp was still prepared on pegs spaced along a wall. But by the middle of the 18th century, the warp was often made on machines very similar to this one.*

The spinster of Fig. 2, on the other hand, is making thread on a very old-fashioned flyer wheel. It was a type that had largely disappeared, except in France. There the peasant women preferred its action. The wheel is turned, not by a treadle, but by a knob on its rim. This leaves only the spinster's left hand free to pull out the unspun fibres. For this reason, the wheel was best suited to flax or hemp, which had long fibres.

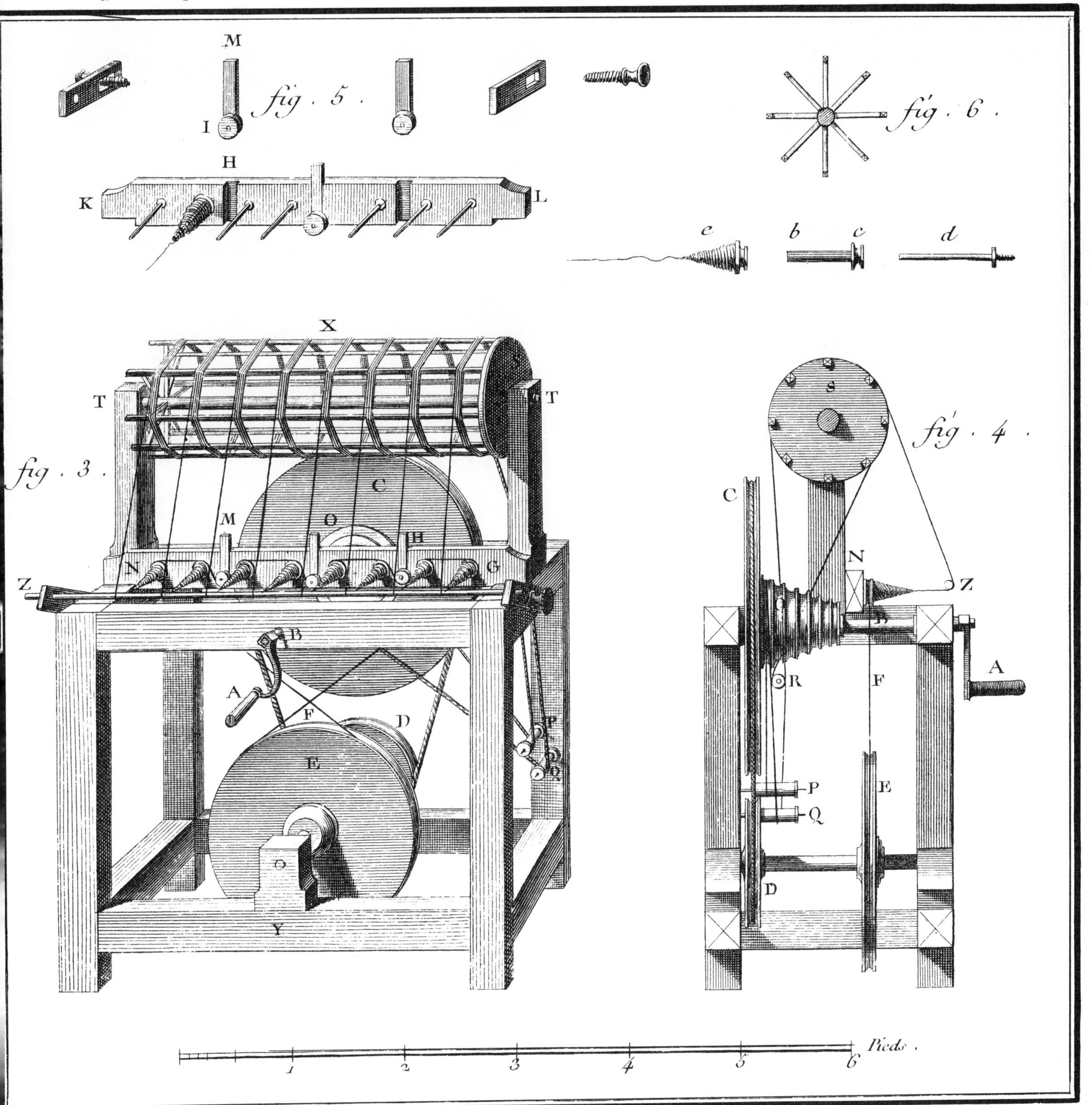
M
fig . 5 .
I
H
K
L
fig . 6 .
e
b
c
d
X
T
T
fig . 3 .
C
M
O
H
N
G
Z
B
A
F
D
E
P
Q
Y
S
fig . 4 .
C
N
Z
A
R
F
P
E
Q
D
1
2
3
4
5
6
Pieds .

Vol. XI, Tisserand, Pl. I.

Plate 309 Weaving I

Weaving is the art of uniting threads into fabrics. All plain woven cloth has the same structure. It consists of a number of longitudinal threads which run parallel to each other and which are crisscrossed and intersected by continuous threads. This "weft" thread goes over and under the "warp" threads and is carried back and forth from side to side of the piece of goods. The longitudinal threads are known as "warp" because they have to be "warped" or pulled taut across a frame. Pattern was achieved by varying the color of the weft or the route by which it was led through the weft. This last was accomplished by raising different sequences of warp-threads for the passage of the shuttle which carried the weft.

The flying shuttle was not in general use when the Encyclopedia *was published. Most ordinary cloth was woven on the horizontal frame loom illustrated in this plate. On this the alternate warp-threads were raised by heddles (or healds) operated by treadles. The shuttle was passed by hand from one side of the loom to the other.*

Plate 310 Weaving II

With this loom the textile industry took its first step into the industrial revolution. It is equipped with John Kay's flying shuttle (Figs. 15 h; 16) which he had patented in 1733. The Encyclopedists urged its advantages upon French industry. In this case little urging was needed.

The loom is powered by footpedals (y). These move the heddles (R, R) which raise or lower alternate threads of the warp. The weaver makes his shuttle fly by jerking the cord (l) with its handpiece in the center. The cord pulls the picker (Fig. 15, g) which strikes the shuttle such a blow that it rolls on its small wheels right across the batten (Fig. 14 g; 15 i) or track prepared for it. When it reaches the side, it slides into the groove formed by the escapement (Fig. 15 k). This arrests the shuttle without rebound. Solving the problem of rebound was the essential trick in Kay's famous device. A second pull on the cord sends the shuttle back. On each passage it moves through the warp along a channel opened by the heddles leaving a trail of thread to form the weft. The shuttle carries a bobbin (Fig. 16 b) and the weft thread pays around the little lateral pulley wheel (Fig. 16 t).

The flying shuttle revolutionized weaving. It increased speed. It cut down manpower. Most important, it permitted a wider web. Thirty inches was the maximum width that a single man could weave on the old loom. Anything wider required two weavers to throw the shuttle one to another.

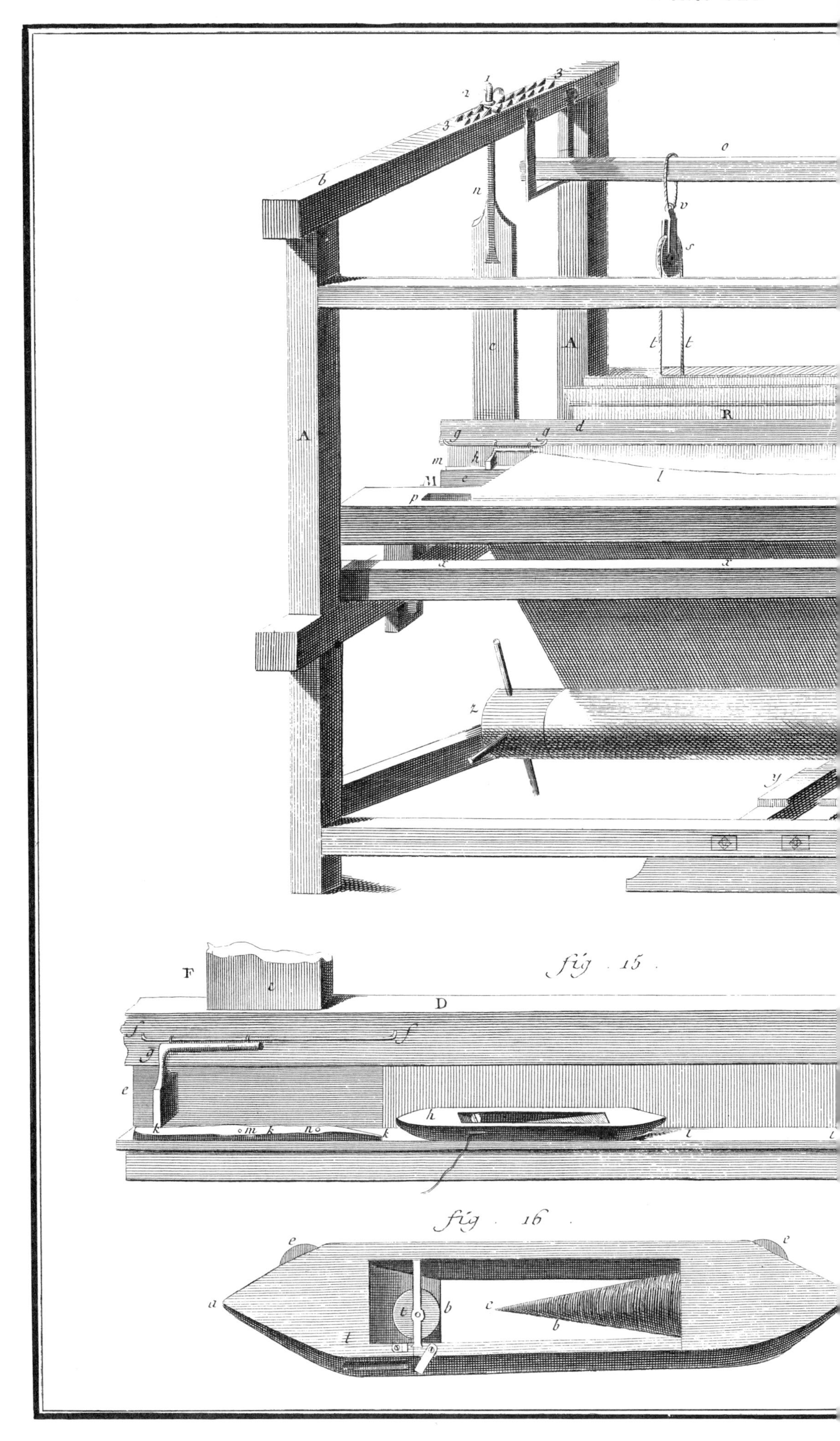
fig. 15.
fig. 16.

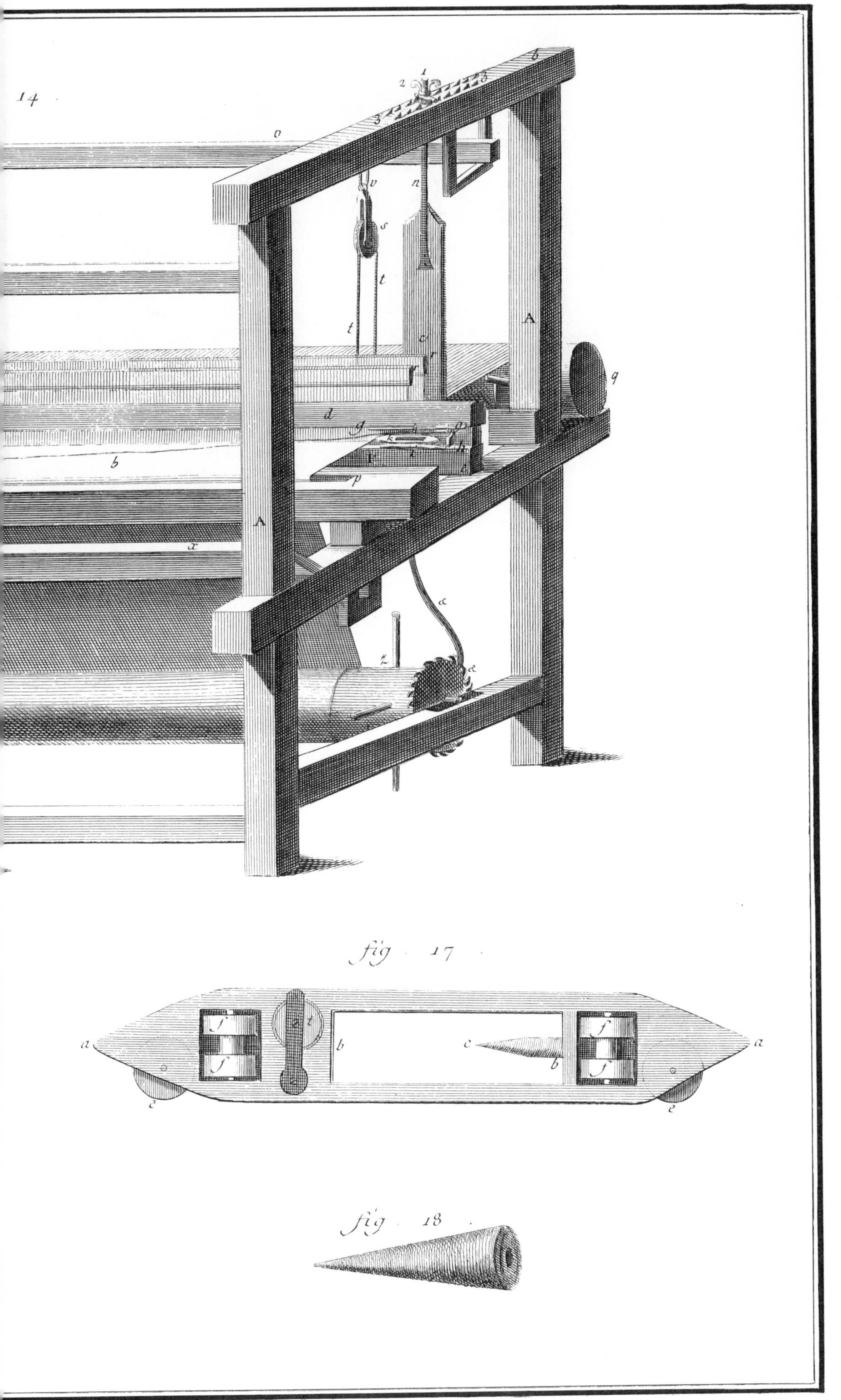
14
fig. 17
fig. 18

Vol. II, Bonnetier de la foule, Pl. I.

Plate 311 Fulling

In the pre-industrial woolen trade the master-hosier put out the weaving of cloth to domestic workers, but he generally fulled and finished the cloth in his own establishment.

The man at the tub (Fig. 1) is a fuller. His work consists of washing and carding greasy impurities out of woolen cloth. He runs hot water into his tub (out of charming faucets), adds soap and fuller's earth as detergents, rubs his cloth against a washboard made of blunt nails, rinses, and wrings out the wet cloth. Since stockings are to be the product, workmen place the wet cloth on stocking forms to give it the right shape.

The other workmen are busy on other products. One cards a piece of cloth for a cap very like the one he wears himself, and the other shears the long fibres from a bolt of serge to give it the even and smooth texture of finished cloth.

Vol. III, Draperie, Pl. VII.

Plate 312 Raising the Nap

After washing and fulling, finished cloths are brushed with teazels to raise the nap.

Plate 313 Combing and Finishing

After the nap was raised, the cloth was combed. Wool-combers (Fig. 39) use a comb with very long teeth. With this they align the fibers of the nap and feel out lumps, knots, and other imperfections.

The tub above the frame (right) is for washing and wringing woolens (Fig. 43); the little spinning wheel (Fig. 45) reclaims fibers combed out of cloth and spins them into new yarn. The device at the bottom (Fig. 46) was peculiar to French finishing processes. It passes the cloth over a charcoal fire (e) so that the reverse side which faces up can be sprinkled with a solution of gum arabic.

These are the last steps in finishing. Now cloth is ready for the tailor or dyer.

fig. 40.
fig. 41.
fig. 42.
fig. 43.
fig. 44.
fig. 45.
fig. 46.
A
B
C
D
E
G
a
b
c
d
e
f
g
h
i
l
m
n
o

Vol. II, Faiseur de métier à bas. Pl. I.

Plate 314 The Stocking Frame

Stockings covered a much larger portion of the 18th century male leg than of its 20th century descendant, which hides its contours in trousers. Stockings might be cotton, wool, or silk according to quality, and manufactured either by hand-knitting or by the stocking frame, one of the proto-industrial machines of the textile trades.

As with many devices, the identity of the inventor is unknown. The machine first appeared in England, but (according to the French legend) did so as the brainchild of a French artisan driven thither by frustration over the ingratitude of his own government. This is a standard story, however, a variant of which attaches to many famous inventions. It is probable that its basis is not so much fact as the fixed French belief that ingenuity springs eternal in French breasts where it is as eternally stifled by failure of the French state to recognize and nurture it.

At any rate this frame (Fig. 3) knits silk stockings. The knitter's female assistant—probably his wife—winds hanks of silk thread onto the special bobbins used in the frame.

Plate 315 Silk I

Vol. XI, Soierie, Pl. I.

Plate 315 Silk I

In its elegance and luxury silk is a fabric which, though it originated thousands of years ago in China, seems particularly suited to the Latin industrial temperament. In modern times France has become the foremost silk producer. The industry was already well seated in southern France in the 18th century, as a result of Colbert's encouragement and the premiums he established for cultivation of the mulberry. But France had not then displaced Italy. It was in northern Italy that the most accomplished silk manufacturers were found.

Silk is unique among fibers in that it comes to the manufacturer already spun—by the worm. It needs only to be carefully reeled. This is the occupation of these two young women. Silk was usually reeled during the months of June and July. After the chrysalides were neatly killed (see Plate 15), the cocoons would be sorted according to size and quality of fiber, and immersed in a hot solution of dilute alkali to soften the gum. Reeling consisted of winding together four or five filaments to make a strand. The girls would join on new filaments as each cocoon gave out.

Plate 316 Silk II

Reeled silk was too fine for most purposes. The process of twisting and doubling, which strengthened silk without kinking it was known as throwing. In France silk was often thrown by hand on a series of reels, wall-frames, and spools, a method which aroused the impatience of the author of the Encyclopedia's *very detailed account* of *the industry. In its stead, he urged mechanization according to the plans printed on this and the following plate.*

This "Piedmont mill" was used in the most progressive factories of Turin. The plate shows the power plant (below) and two great cages (above), one of which fitted inside the other. The inner frame (Fig. 1) transmits its rotation to the spools and spindles of the stationary outer frame (Fig. 2) by means of friction exerted through the curved strips attached to the central post.

Plate 317 Silk III

On the floor above the spindles are these mechanized spools, which wind up hanks of silk thrown in the course of the operation. The whole factory was run on a series of take-offs from a single power source—an illustration which demonstrates that the new element in the "factory system" in England, a little later, was one of scale, not principle.

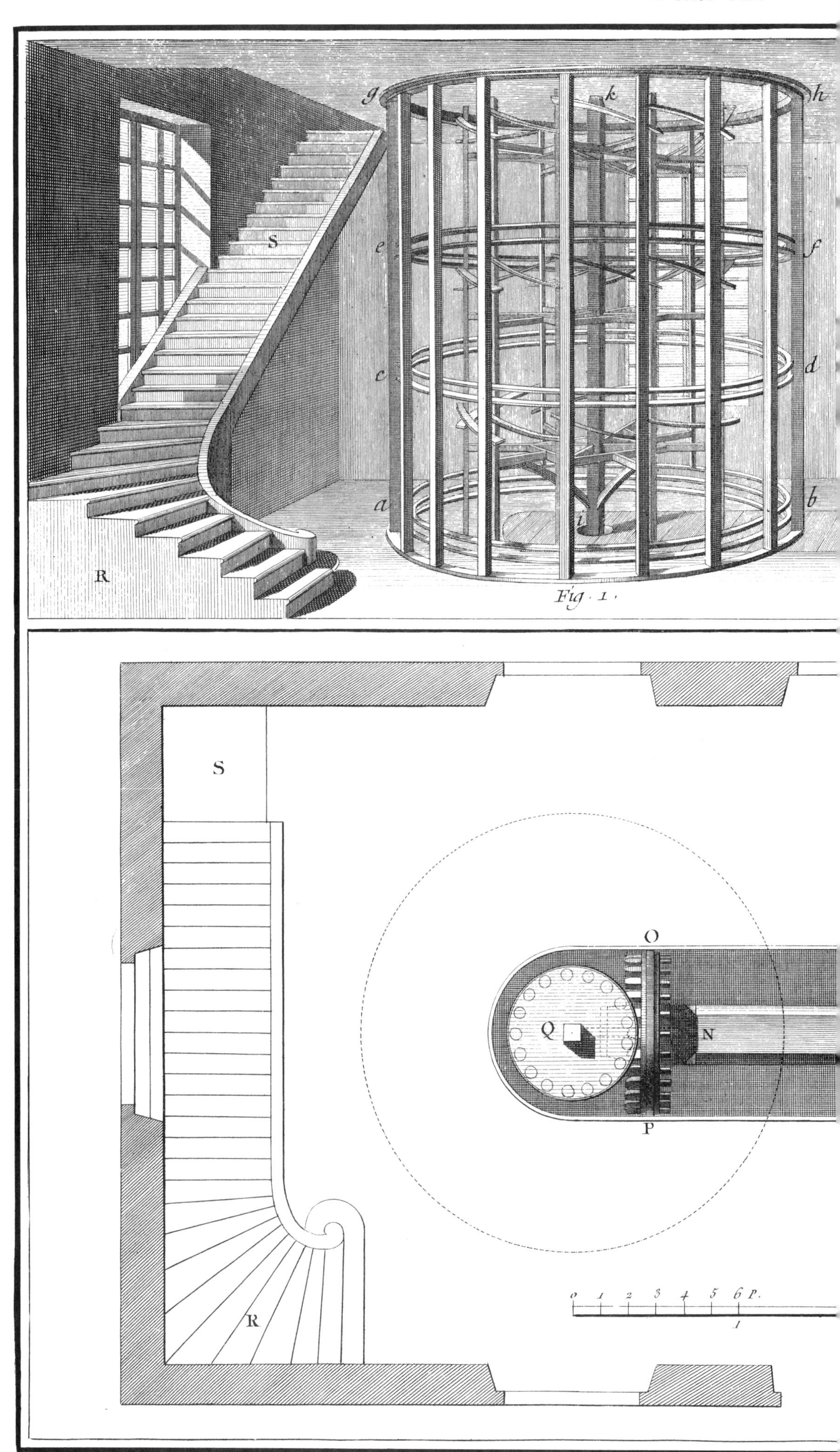
g
k
h
S
e
f
c
d
a
i
b
R
Fig. 1.
S
O
Q
N
P
R
0 1 2 3 4 5 6 P.
1

k
h
f
d
b
B
A
Fig. 2.
B
G
E
BB
L
K
I
D
C
F
H
AA
Toises.
3
4

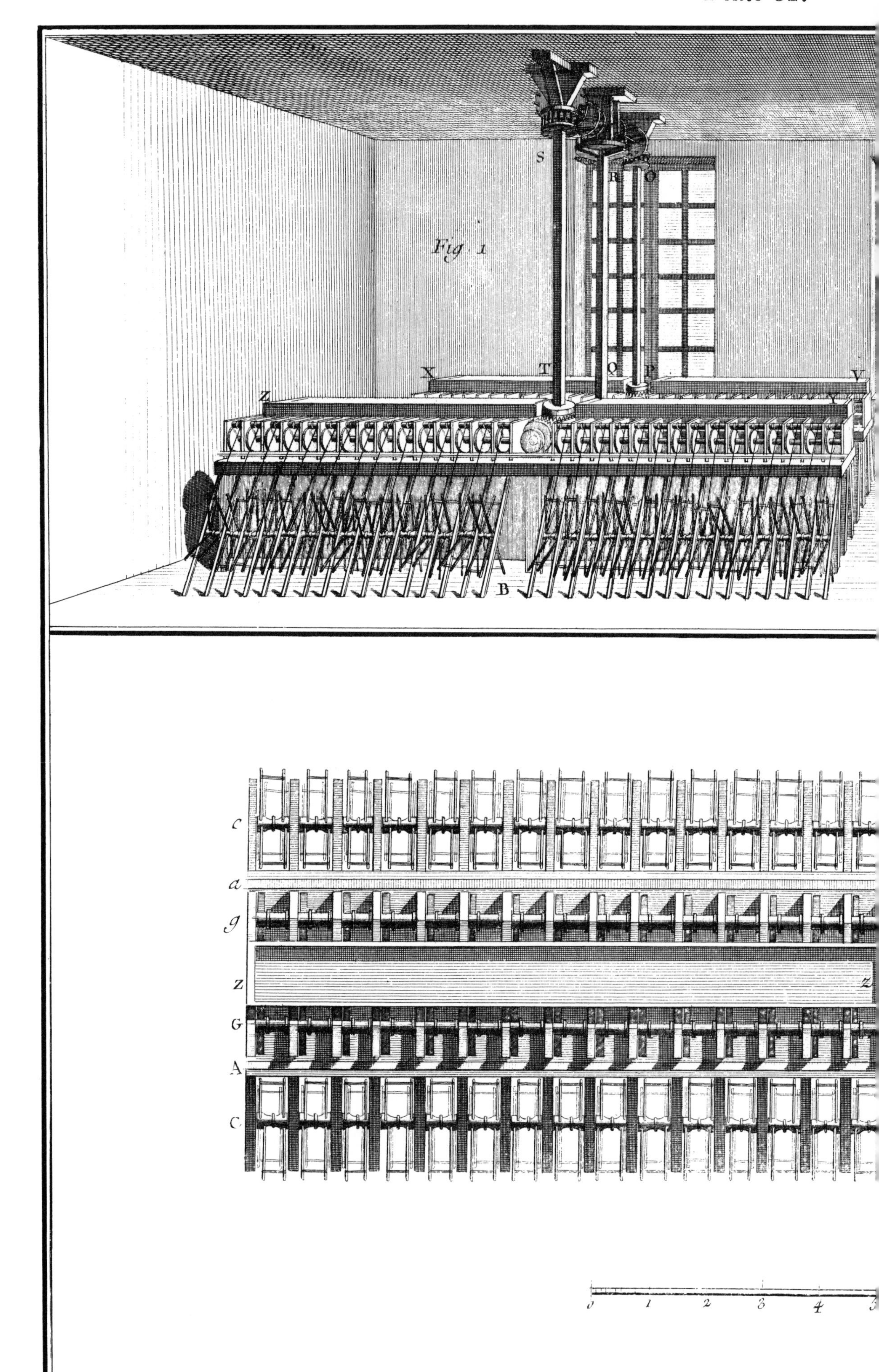
Fig. 1
S
R
O
X
T
O
P
V
Z
Y
B
c
a
g
z
z
G
A
C
0
1
2
3
4
5

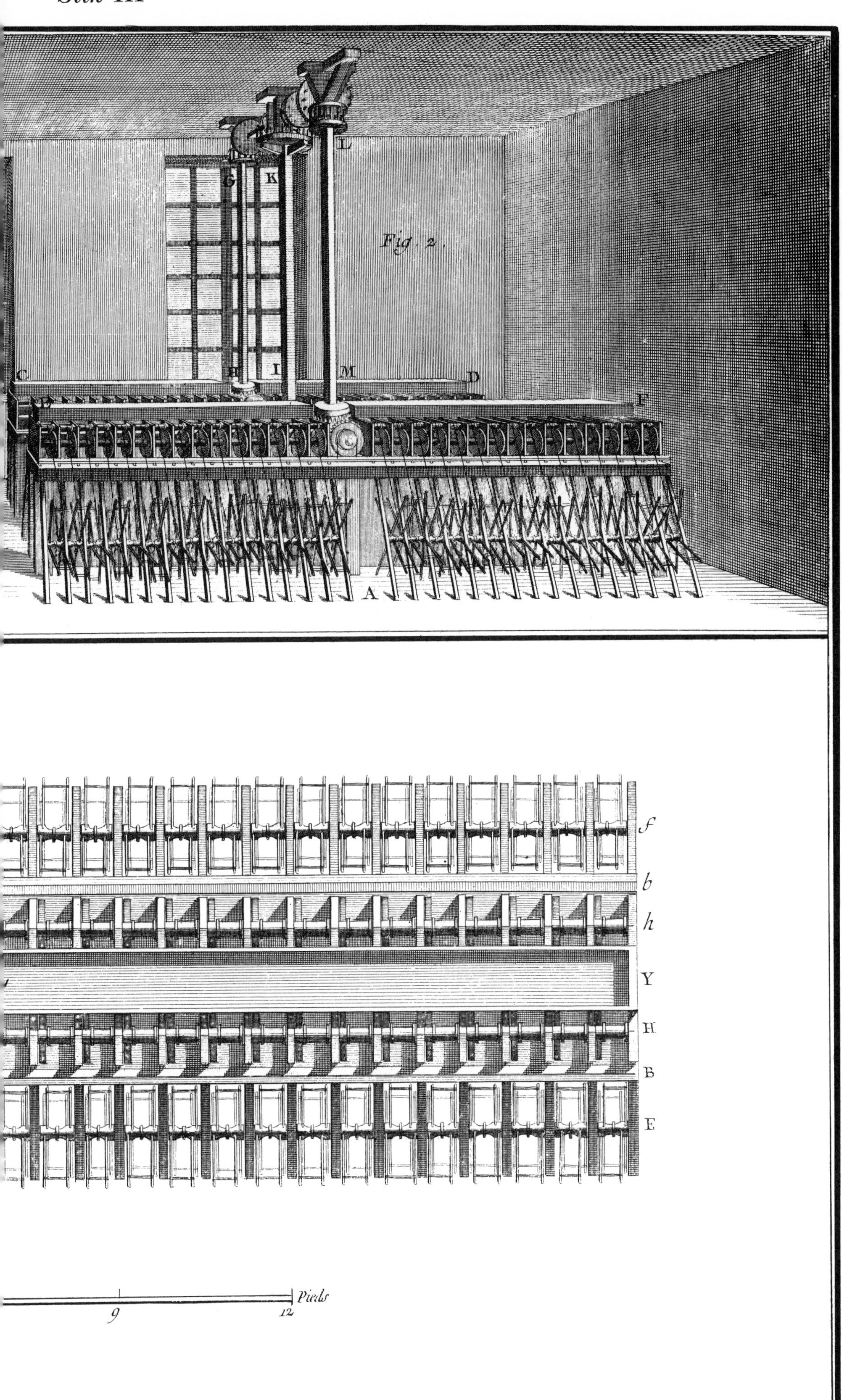

Fig. 2.
L
G
K
C
H
I
M
D
F
A
f
b
h
Y
H
B
E
Pieds
9
12

Vol. XI, Soierie, Pl. XXIII

Plate 318 Silk IV

After being thrown, silk yarn was handled much like that of other substances. The next three plates are concerned with preparation of warp-thread for the loom.

Vol. XI, Soierie, Pl. XXVIII.

Plate 319 Silk V

Warped yarn is drawn from the reel onto a spool.

Vol. XI, Soierie, Pl. XXX.

Plate 320 Silk VI

And from the spool (A) onto a roller, which will stretch it on the loom.

Plate 321 Silk VII

After being woven, silken cloth might be processed in various ways to bring out different qualities. This is a shop for "watering" silk. The term is a misnomer, for the technique consisted of crushing the fabric on a marble slab by rolling it around copper rollers (8, 9) weighted down by an enormous mass of masonry (HILK). At the left a workman rolls a bolt of silk around one of the copper cylinders. When the water wheel has moved the great mass to one end of its course, he will slip this roller under it to be rolled back and take out the far one (9) which will have been left free. When the silk is unrolled, it will be seen to have developed a high sheen and distinctive wavy markings.

Z
P
n
g
m
f
D
B
C
A
H
G
Fig. 1.
Fig. 2.

X
Y
V
Q
R
S
O
K
N
L
7
5
6
H
M
I
3
4
8
E
F
11
A
B
T
C
D

Vol. XI, Passementerie, Pl. I.

Plate 322 Passementerie I

There are styles in industry as in everything else. Just as in metallurgy, England was famous for furnace and foundry and France for ornamental ironwork, so in the textile trades France never successfully challenged English supremacy in mass production of cottons or in solid worth of woolens, while England never had the flair for fancy work—ribbon, braid, gimp, lace, belts. The very word passementerie *has no English translation, and the* Encyclopedia's *specification of what came under it suggests in the words themselves the frothiness of these frivolities:* rubans, galons, dentelles à l'oreiller, ou fuseau, à l'épingle, à la main, houppes, bourrelets, campanes, crépines, bourses, tresses, ganses, nates, bracelets, rênes, guides, cordons, chaînes, éguilletes, ceintures, lacets, rézeaux, cordonnets, canetilles, bouillons, frisons, guépiers. *There is a French proverb which runs* Quand on prend du galon on n'en saurait trop prendre. *The literal meaning is "The more braid the better," but the sense is "You can't have too much of a good thing." But whatever the ultimate frippery, the first step was warping the thread, usually of silk, but for some purposes of fine cotton or linen, and sometimes of gold.*

Vol. XI, Passementerie, Pl. III.

Plate 323 Passementerie II

Plaiting fancy cord was no different in principle from making rope. One could lay the strands in a large number of different designs depending on the pattern and intended use, whether for gilded epaulets, for button frogs, for adorning petticoats, or whatever. One workman (C) is warping alone with a hand wheel—the others are using a large wheel and a movable frame.

Vol. XI, Passementerie, Pl. XIV

Plate 324 Passementerie III

Braid for military uniforms would be of mingled silk and gold thread (b, c), and the brilliance of the gold could be much enhanced by smoking the braid (d) over a brazier of burning partridge feathers and scarlet dye-stuff. The smoking had to take place in a tightly sealed room, and it was not considered an altogether reputable practice, for the sheen it lent to gold braid did not last.

Vol. XI, Passementerie, Pl. XV.

Plate 325 Passementerie IV

Braid implies fringe to set it off, which might be crocheted by hand for pillow slips and bed linen (b, c). Or the loom (a) might be adapted to weaving fringe by machinery.

Plate 326 Ribbons & Ornaments

For decorating ladies' gowns, hats, or lingerie the ribbon most commonly used was made of multicolored silk and called nonpareille, a term still occasionally used in both French and English. It consisted ordinarily of twenty silken strands, each composed of sixty threads. Its method of manufacture was curious, for instead of being woven, it was rolled and welded almost like a strip of metal.

The mill is illustrated opposite, in Fig. 1. The upper roller was made of wood and the lower of copper, so that it could be heated over charcoal before being used. The silken strands were fed between the rollers, where the heat partly fused the threads together. To make the union permanent nonpareille was then sized with glue made of old parchment, after which it was wound onto a drying spool.

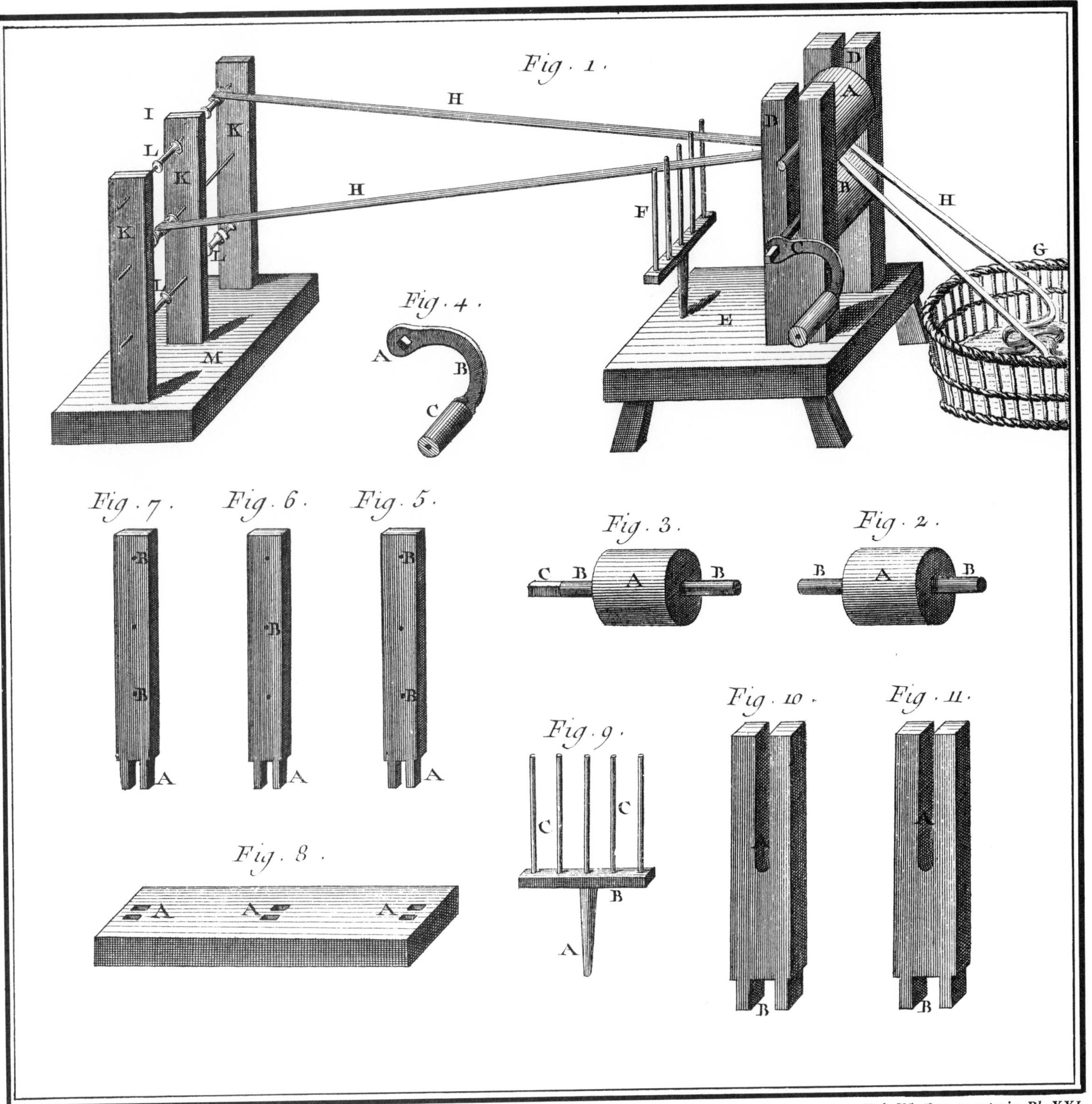
Fig. 1.
H
I
K
L
K
H
K
L
L
M
D
A
B
B
F
C
E
H
G
Fig. 4.
A
B
C
Fig. 7.
B
B
A
Fig. 6.
B
A
Fig. 5.
B
B
A
Fig. 3.
C
B
A
B
Fig. 2.
B
A
B
Fig. 10.
A
B
Fig. 11.
A
B
Fig. 9.
C
C
B
A
Fig. 8.
A
A
A

Plate 327 The Drawloom I

Figured fabrics in which a design is repeated at regular intervals were woven on a special loom, the drawloom. It employed two harnesses for manipulating the warp (g, g). The larger (P) is the heddle-harness. Its cords are worked by treadles and pass over a pulley frame (F). The heddles raise warp-threads for the binding weave, or the passages of the shuttle which make the background, not the pattern.

The figure-harness (S, S) is the distinctive feature of the drawloom. It owed its invention to the impossibility of crowding a large number of heddles into the width of the weave. This limited the broadloom to the simplest designs. The essential part of the figure-harness is the comber board at the top (upper S). This is a perforated board through the holes of which vertical neck-cords pass to be attached to the warp-threads (g) by a metal eye coupling, the mail. The neck-cords are weighted by wire lingoes (lower S) which pull the warp-threads back into position after each "pick" of the shuttle. After passing through the comber board, a number of neck-cords (usually four) were tied to a single tail-cord. The tail-cords (f) are lifted by the vertical simples which in this case appear to be attached to heddles—though the cords which drew the simples were called lashes rather than heddles.

It may clarify the operation of the loom to run through the motions it makes for a single passage of the shuttle. The plate shows the first lift of the harness in the weaving of a figure of the pattern. The weaver presses the appropriate treadle with his foot. The harness-cords or lashes attached to it over the pulley frame raise the simples. These are tied into the tail-cords in a prearranged pattern. Each tail-cord raises four or more neck-cords in the figure-harness. These in turn raise those warp-threads which make the appropriate shed for the particular design-pick or passage of the shuttle. The weaver then beats the weft into the fabric with a reed or comb. The lingoes pull the warp back into position, and the apronlike weight on the lower end of the simples keeps the cords of the heddle-harness under tension.

The principle was that combinations of lashes, each acting through tail-cords and comber board on a number of warp-threads, were substituted for the individually acting heddles of the broadloom. The pattern to be woven would be tied into the harness before weaving, much as "information" is read into a modern calculating machine. Each successive set of the simples would lift a preordained combination of warp-threads. The pattern could be repeated as often as necessary.

G
H
H
F
I
K
L
D
E
D
Q
S
R
R
R
P
R
O
A
A
g
k
l
m
C
C
M
S
N
N
1
2

F
G
K
I
L
E
A
T
f
u
Y
e
d
X
X
q
c
A
h
U
g
a
b
B
m
r
C
Y
C
Pieds
4
5
6

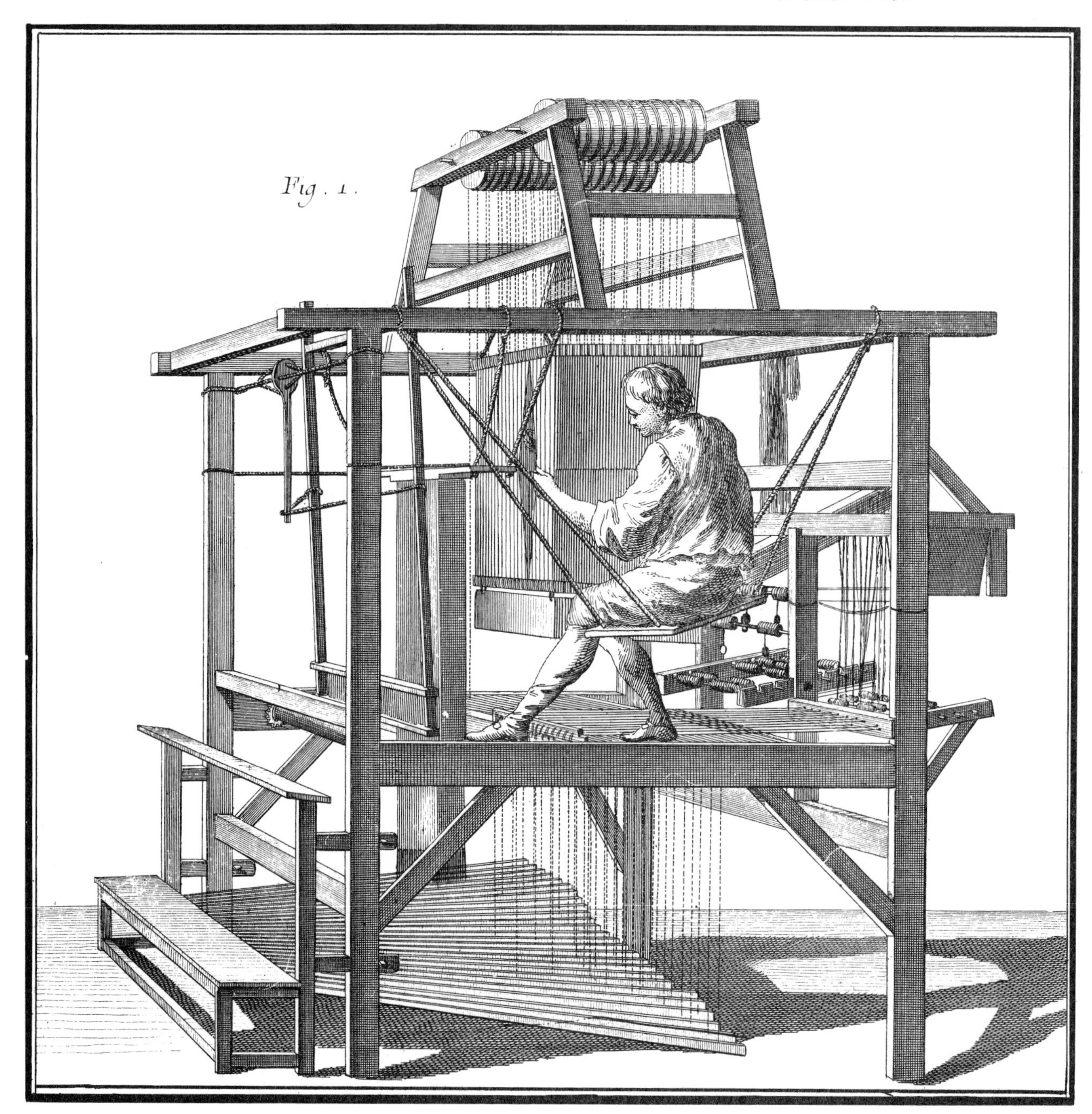

Plate 328 The Drawloom II

The weaver himself tied up the pattern he was to reproduce. Fig. 2 is a portion of the finished weave. He was given the design on sheets of ruled paper, "point-paper" (Fig. 3). Since the pattern is symmetrical, he needs only half of it. The lower half will be a simple repeat in reverse order through the figure-harness. Each individual cord of the simple is shown as a small square in the grid which itself is ten squares by twelve. Since this pattern requires eight grids, each of which is ten squares across, the loom has to carry a width of 80 lashes.

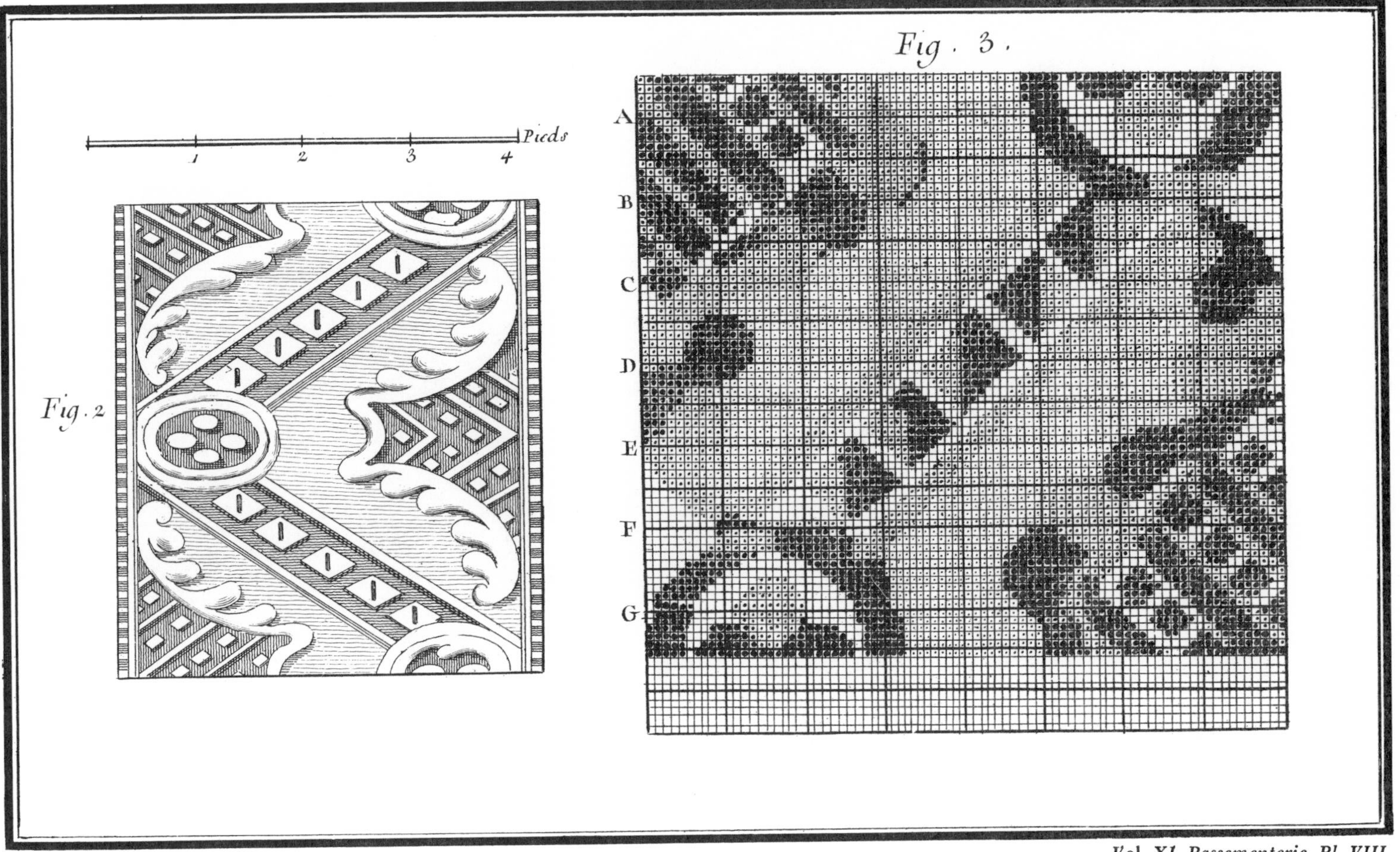

Vol. XI, Passementerie, Pl. VIII.

Plate 329 Gobelins Tapestries I

In the 18th century there still flowed through the Quartier Saint-Marceau a small tributary of the Seine called La Bièvre. It continues to run underneath the 5th Arrondissement, but it has been entombed in sewers, and its existence has been forgotten along with its name. On its bank stood one of the world's most famous concerns, the home of Gobelins tapestries. It had been created by Colbert as a royal manufactory in 1667 on the site of an old dyeworks established two centuries earlier by Jean Gobelin, who had been the first in France to have the secret of a true scarlet.

The history of tapestries can be traced all the way back to the ancient world until, with the story of Penelope's web, it turns into myth. The weaving of tapestry was one of the earliest arts to revive in medieval Europe. Many critics feel that it reached its highest pitch of artistic excellence in the 14th and 15th century tapestries of France and Flanders. In modern centuries France has proved the most congenial environment, in part because of the official leadership of the Gobelins factory.

The Gobelins owed its foundation to that same impulse which led the French state, la grande nation, *to take under its protection letters, art, and science as essential embellishments of a great monarchy. It is to this patronage that France, and indeed the world, are indebted for the Comédie Française, the Louvre, the gardens of Versailles, the Académie Francaise, and the Académie des Sciences.*

Despite this official support, the 17th and 18th centuries do not now seem the happiest periods in the artistic history of tapestries. The medium is best suited, perhaps, to styles which, like the medieval or the contemporary, are highly stylized or abstract.

Neither the baroque magnificence of the 17th century, nor the pastoral and somewhat effeminate prettiness of the 18th, is pleasing to modern taste. Nevertheless, a comparison might be made between what the existence of the Comédie Française has meant to the French theater, and what the Gobelins factory has meant to tapestry weaving. The official styles are vulnerable to the criticism of academicism. Nevertheless, the disinterested support of these arts by the French state has provided at once that basis of security and that prick towards rebellion which insure that the adventurous spirit, irritated into independence by officialdom, will not risk starvation through excess daring. There is a fold for the rebel's return.

The Gobelins has always been a mecca for tourists. The tableau shows one of the haute lisse *or high warp galleries. In this method the warp is stretched vertically before the weaver. What first strikes the visitor is that the weaver—there is one peeking out between the threads near the fleur de lys on the right-hand frame—never sees the front of his tapestry. He has to follow the design by working from the reverse side.*

As the work progresses, the tapestry is rolled onto the lower beam or roller. Wooden guards (c) protect it from damage or dirt. Weft threads in the different colors are wound onto bobbins, an armful of which are being carried to a weaver by an attendant (g). Another weaver searches out the exact shade he needs in his storage chest (m). Dress is always an index to social esteem, and that of the weavers will suggest that theirs was no menial profession. They are the aristocrats of textile workers, artists rather than artisans. A little farther on another weaver (c) has come out front to pick a bit of fluff or a dustroll off his warp. And in the center a spinster works her wheel, served by a boy with an armful of woolen hanks.

Fig. 1.
q
q
q
l
p
m
n
f
f
f
i
h

Vol. IX, Tapisserie de Haute Lisse des Gobelins, Pl. I.

Plate 330 Gobelins Tapestries II

Tension on the warp was maintained by brakes, one pulled tight around the upper roller beam (Fig. 1) and bound to its mate lashed around the lower (Fig. 3).

Vol. IX, Tapisserie de Haute Lisse des Gobelins, Pl. VI.

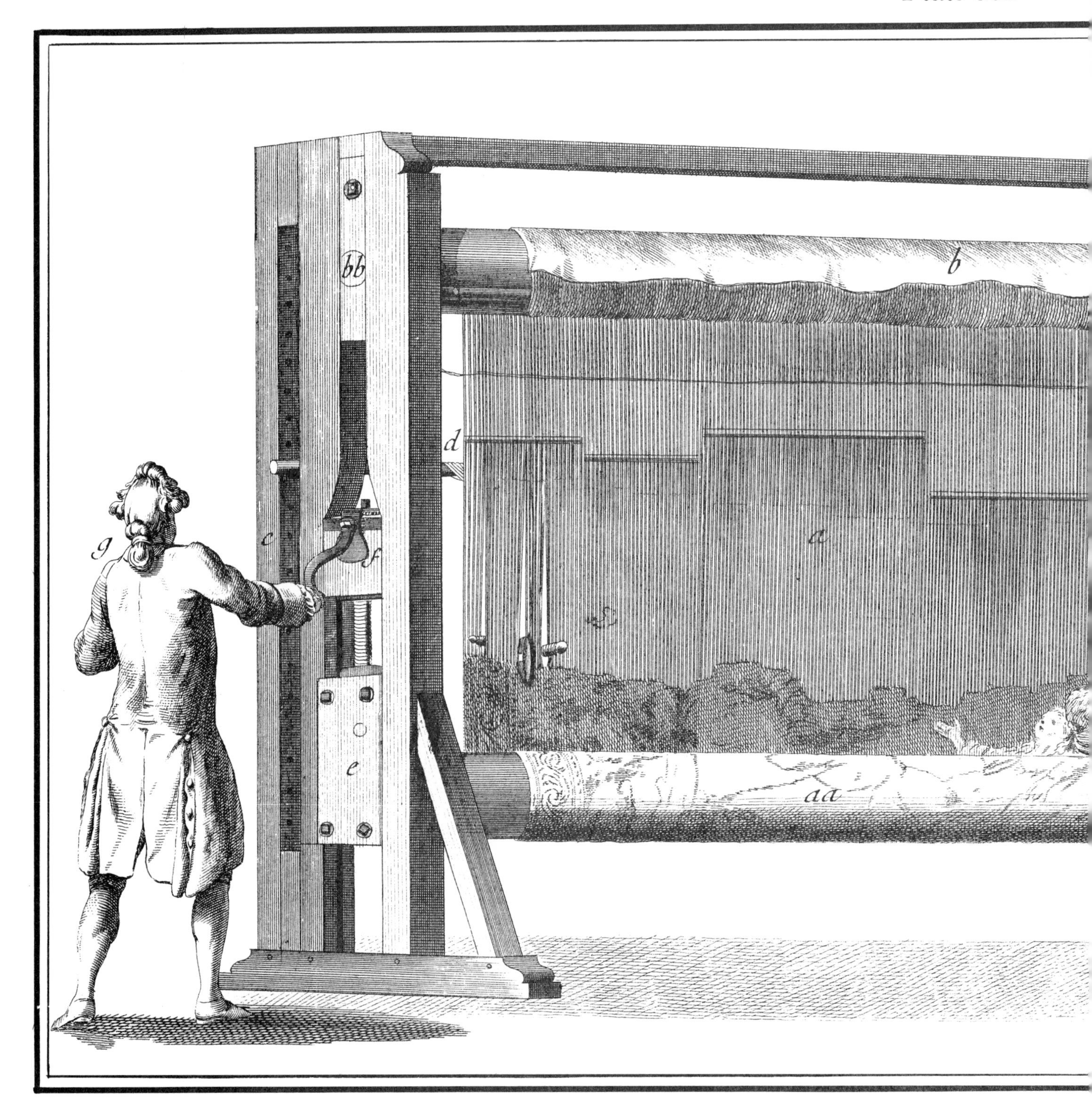

Plate 331 Gobelins Tapestries III

A new invention for tautening the warp was this screw arrangement (ff). It required less service and was not likely to let go (as the brakes might do), endangering the weavers. Notice from the glimpse of the work how it appears to the weaver, as he looks at the wrong side while constantly rolling out of sight what he has done.

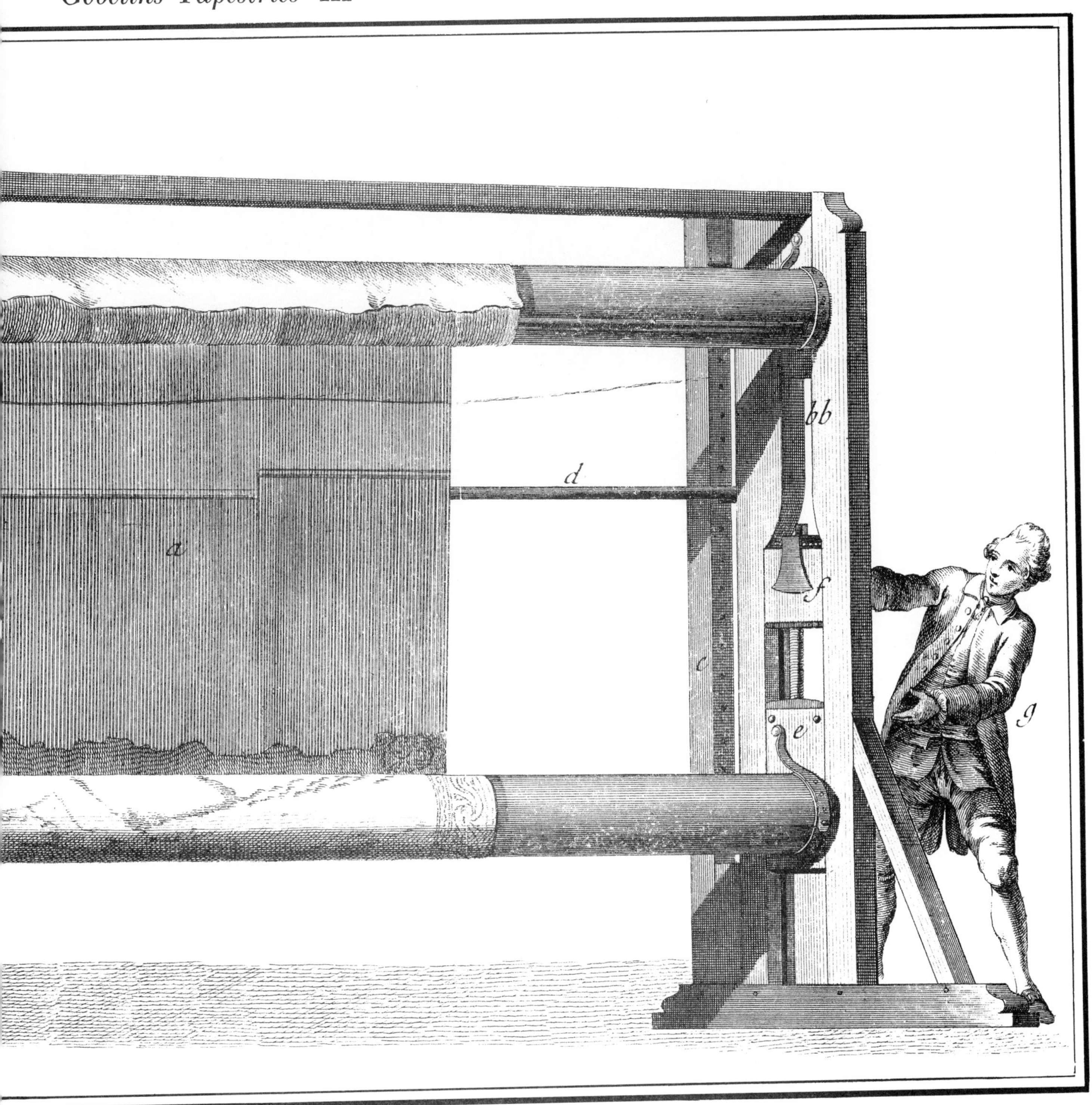

Vol. IX, Tapisserie de Haute Lisse des Gobelins, Pl. VII.

Vol. IX, Tapisserie de Haute Lisse des Gobelins, Pl. IX.

Plate 332 Gobelins Tapestries IV

The technique of weaving tapestries differs in one essential respect from that of the broad loom. The weft is discontinuous. The threads which compose it are knotted on where required by the pattern and cut loose when their color is to disappear. It is not as in cloth. The weaver does not carry them across concealed behind the fabric.

The weaver holds a bobbin in his right hand. With his left he fingers lashes tied to the warp-threads. With these he pulls out the warp-threads so that he can pass his bobbin behind them. Bobbins of weft carrying thread of various colors hang from the piece at the points where he has last used them.

Plate 333 Gobelins Tapestries V

Fig. 1 is a close-up of a weaver's hands. The left pulls the lashes so as to move selected threads of the warp forward, while the right handles the bobbin of woof. Below and at the left the point of the bobbin is inserted between the threads—illustrating, incidentally, why the bobbin must be pointed.

In certain circumstances—for example during a run of plain color—it was quicker to dispense with the lashes and to handle the warp-threads directly. This is the operation illustrated in Fig. 2.

Plate 334 Gobelins Tapestries VI

Fig. 1 shows how the weaver traces the design on the warp. A scroll-like arrangement bearing the design is passed behind the warp cords to which the artist transfers the outlines by blacking with a piece of graphite. Below in Fig. 2 the weaver beats the weft down into the fabric with a battening comb.

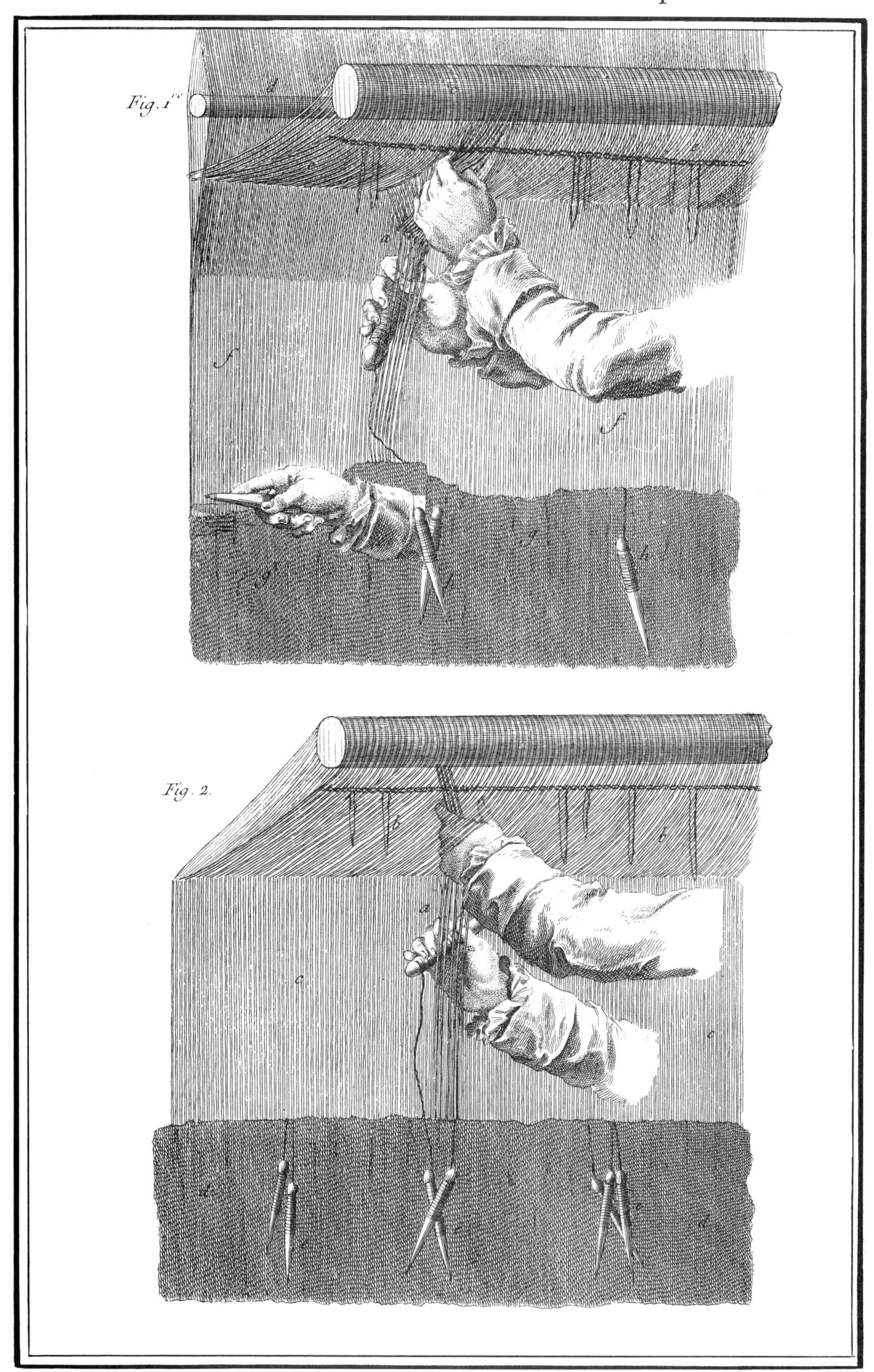
Fig. 1re
d
c
e
a
f
f
g
g
h
Fig. 2.
b
b
a
c
e
d
d

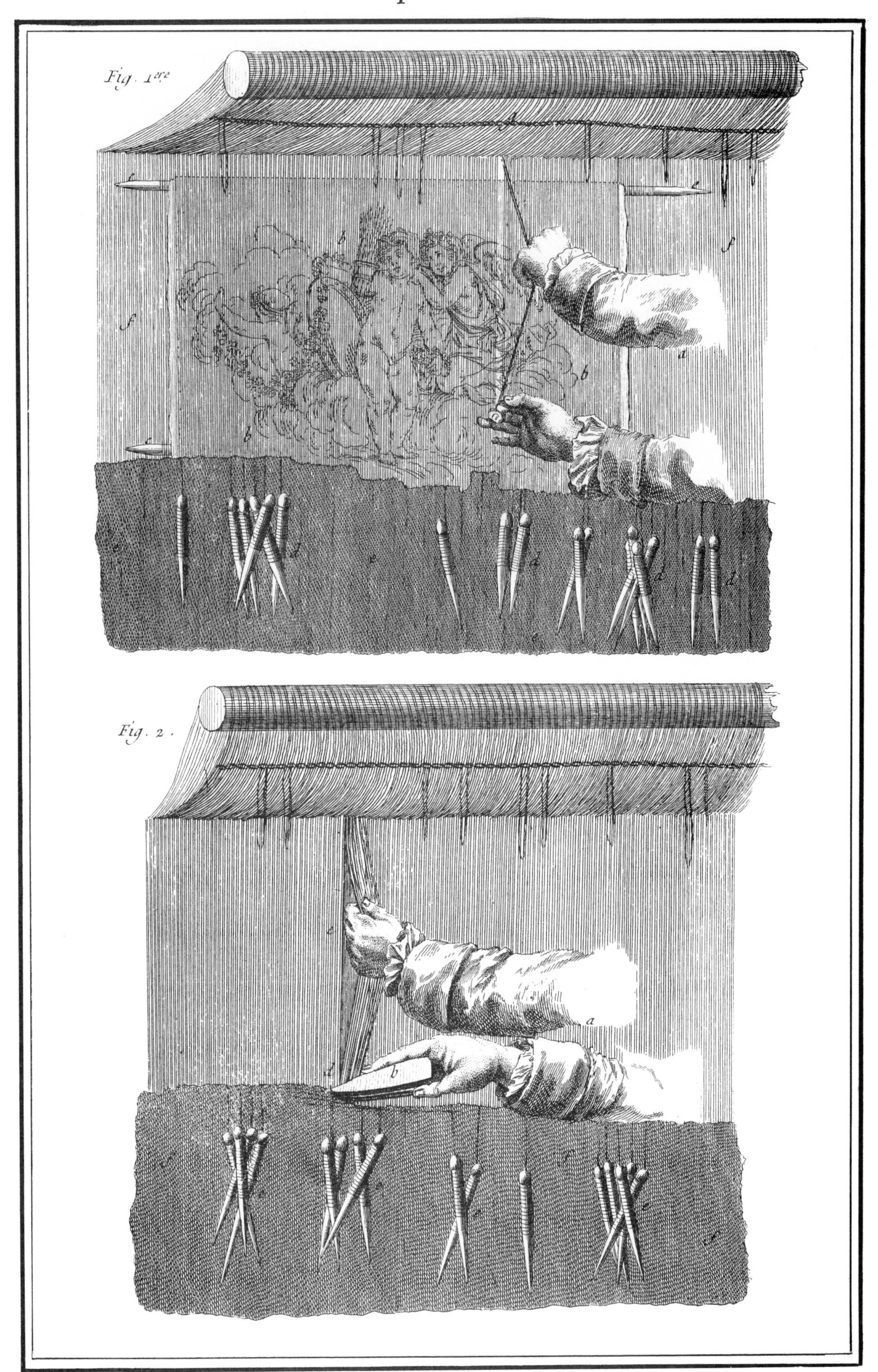

Vol. IX, Tapisserie de Haute Lisse des Gobelins, Pl. XI.

Vol. IX, Tapisserie de Haute Lisse des Gobelins, Pl. XIII

Plate 335 Gobelins Tapestries VII

A half-finished tapestry from the front of the loom. The chains and sticks (b, c, d) running across the warp serve to keep the threads in place and are removed as the weaver reaches them. The roll at the top is protected by a piece of serge (a) and the finished work at the bottom by a wooden rampart. The smaller wooden panel (i) serves to shade the worker's eyes; his judgment is likely to become distorted if he faces the light through his threads the entire day.

Plate 336 Gobelins Tapestries VIII

Basse lisse *or low-loom weaving differed from* haute lisse *in that the warp was stretched (d) horizontally instead of vertically, and instead of sitting up to his work, the weaver bent over it. This is the only difference, however. The method of weaving was essentially the same, as were the tapestries produced.* Basse lisse *was preferred in Beauvais, the second great tapestry center in France, but the Gobelins also contained this gallery equipped for it.*

In the foreground a worker winds a hank of yarn, and a tracer (c) copies a pattern for the weaver's use. The colors on the shelf (f) have fallen into disorder and require sorting out, while at the cupboard (m) an assistant looks for a ball of wool, silk, gold or silver thread needed at the looms. A bar on pulleys (h) serves to hoist a finished piece the height of the room so that it can be viewed and inspected as a whole.

h
m
m
m
f
l
d
a
a
c

Vol. IX, Tapisserie de Basse Lisse des Gobelins, Pl. I.

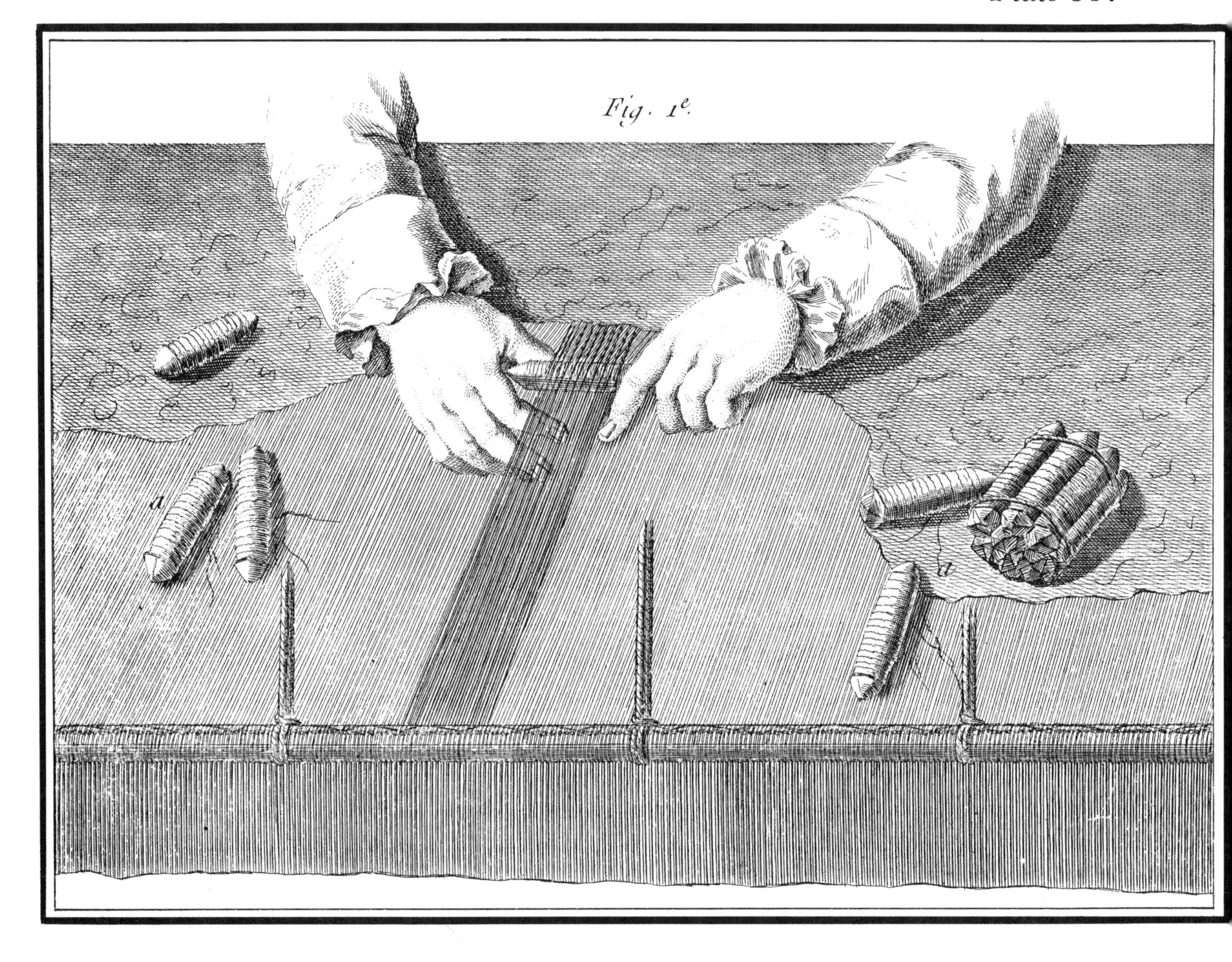

Plate 337 Gobelins Tapestries IX

Weaving on the basse lisse *loom. The bobbins are designed somewhat differently, since instead of hanging by their own threads when out of use, they lie on the surface of the work.*

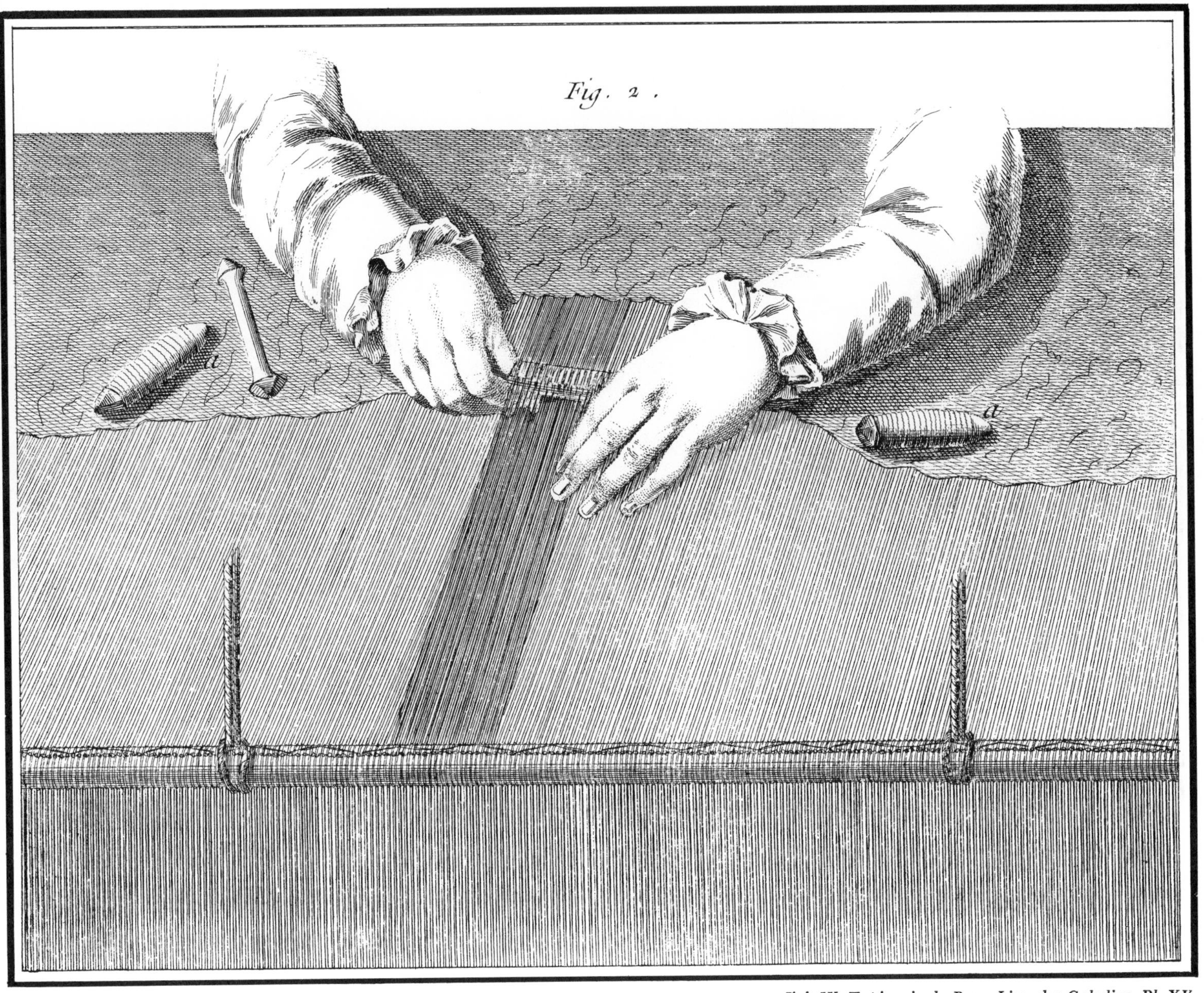

Vol. IX, Tapisserie de Basse Lisse des Gobelins, Pl. XV.

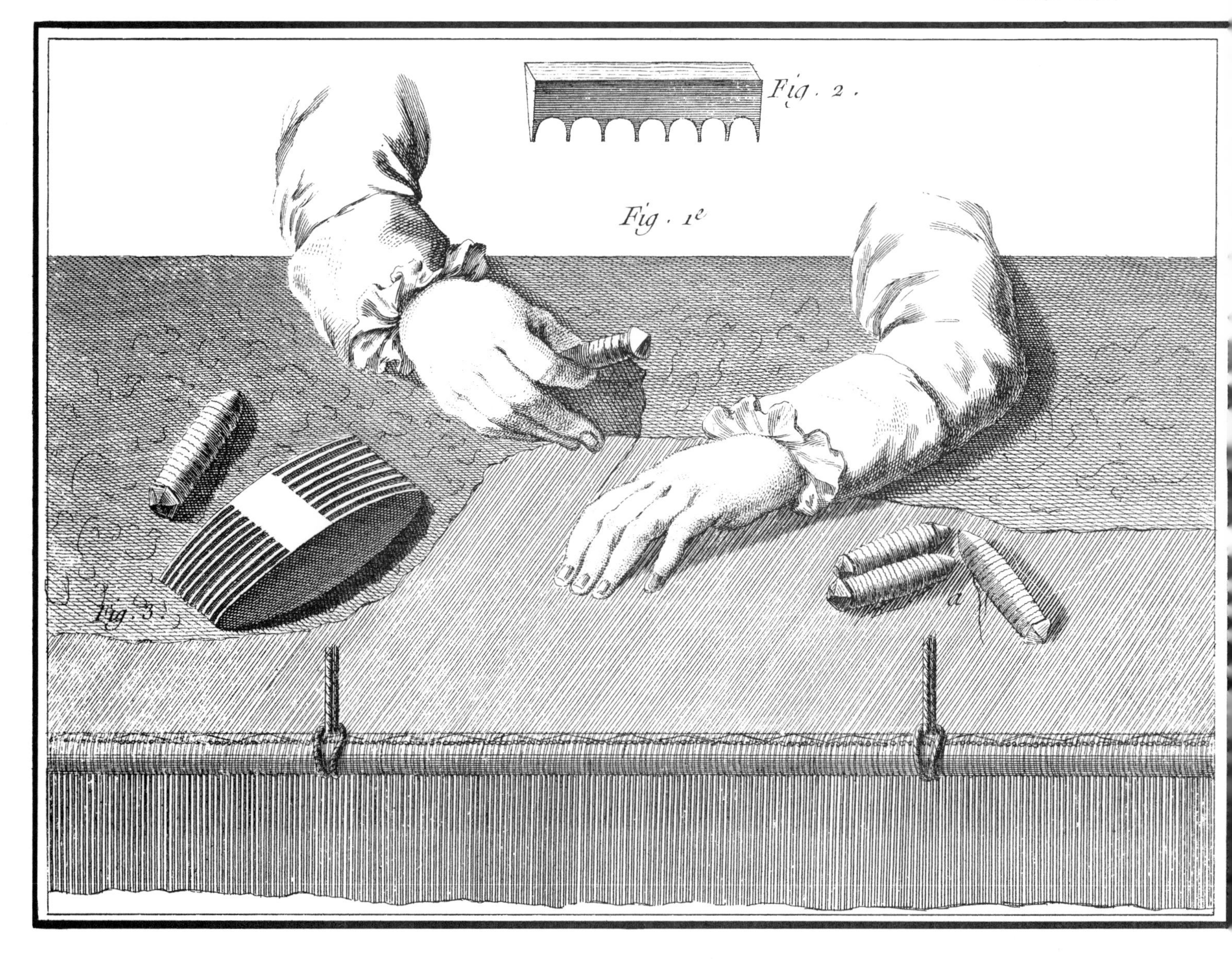

Plate 338 Gobelins Tapestries X

The weaver uses his fingernail to tuck the weft threads down against the woven stuff (Fig. 1) and finishes the compacting with a comb (Fig. 4).

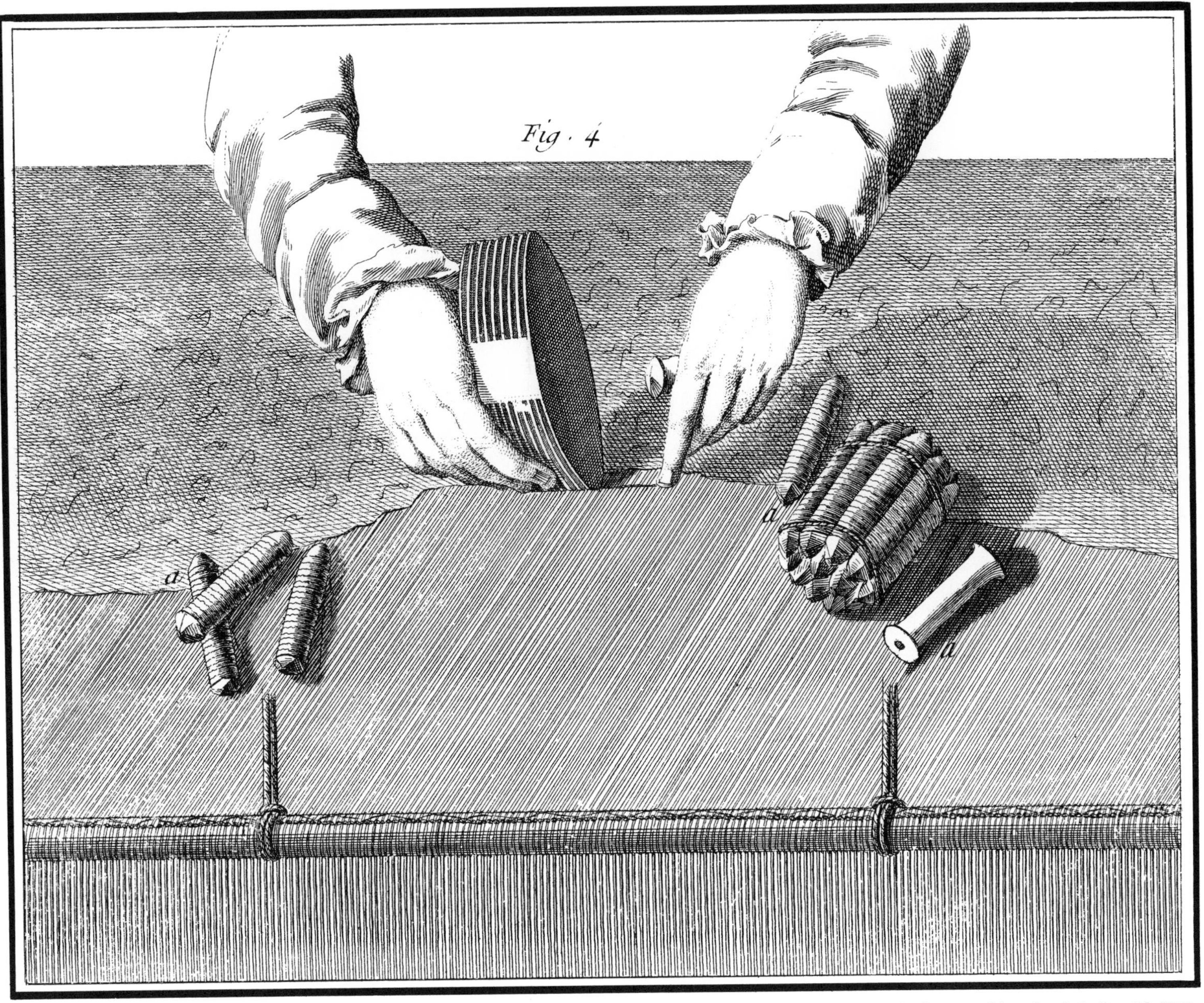

Vol. IX, Tapisserie de Basse Lisse des Gobelins, Pl. XVI.

Plate 339 Gobelins Tapestries XI

The junction between threads of different colors leaves little gaps in the finished weave to be sewn up afterwards.

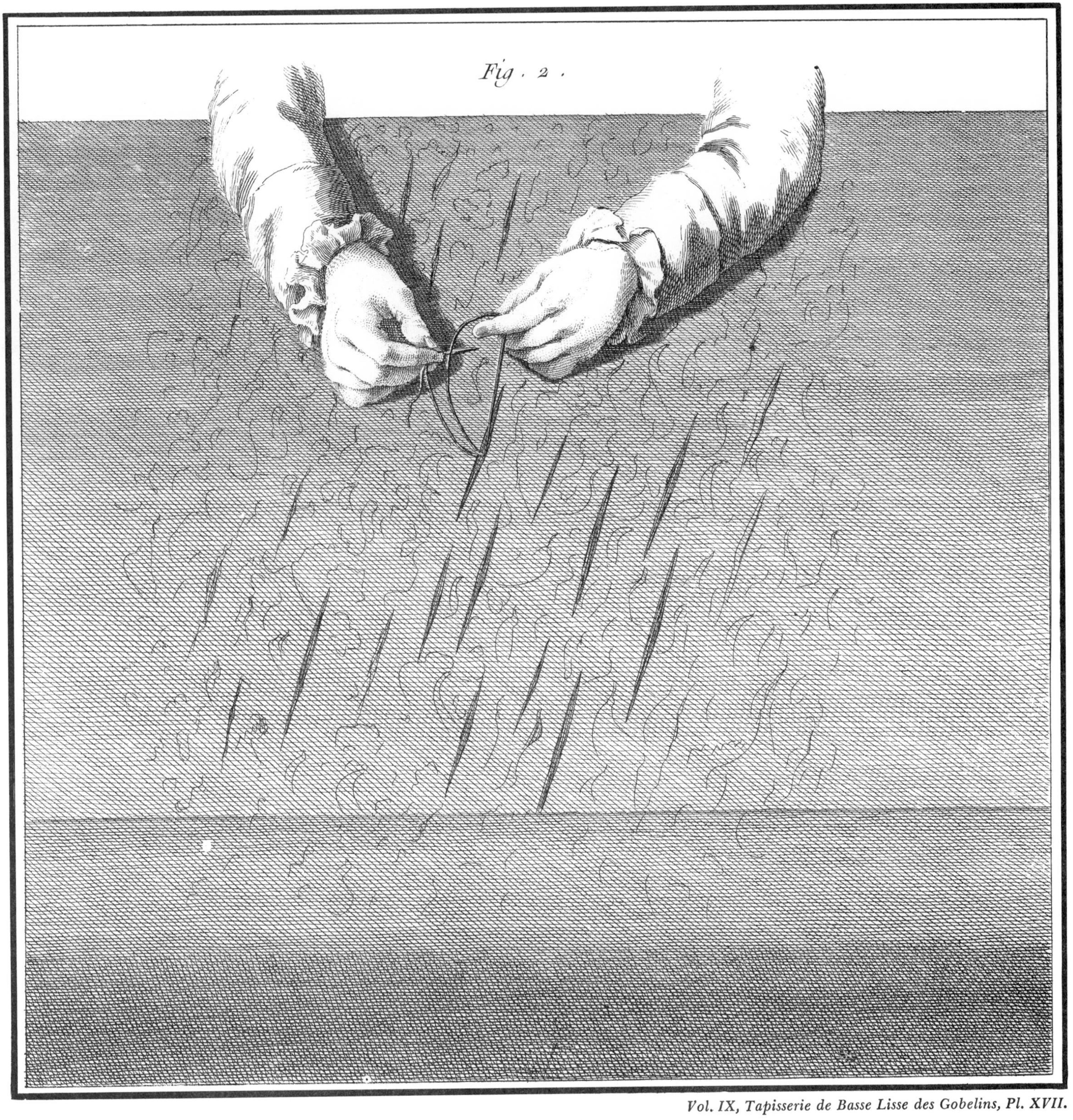

Vol. IX, Tapisserie de Basse Lisse des Gobelins, Pl. XVII.

Plate 340 Gobelins Tapestries XII

A weaver works on into the evening by the light of a candle hooked above his loom. The German bombardments of 1940 destroyed the Beauvais tapestry works, but tapestry weaving (and the Gobelins with it) has undergone something of a renaissance in Paris in recent years. It is still possible to visit the factory, where some of the original looms continue in use on which the weavers work modern designs furnished them by contemporary artists. The weavers themselves are still housed on the premises, though unfortunately the crowding in of Paris has deprived them of the individual gardens allotted by Colbert.

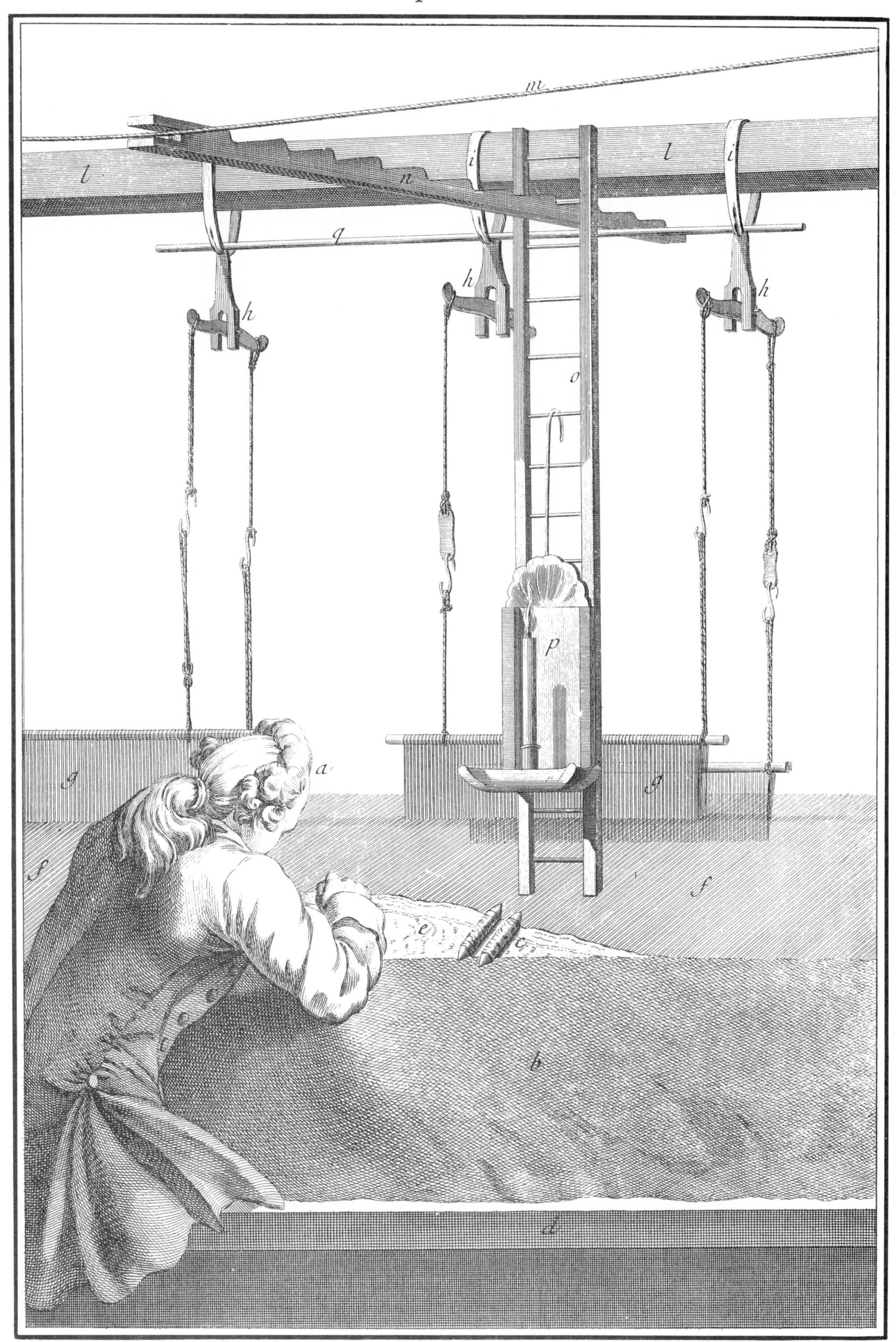

Vol. IX, Tapisserie de Basse Lisse des Gobelins, Pl. XVIII.

Plate 341 Carpets I

Besides the Gobelins, there was a royal manufactory for the weaving of Turkish carpets in the Faubourg de Chaillot. It stood on the site of an old soapworks near what is now the Trocadero. Never having attained the reputation of the Gobelins, it disappeared during the Revolution, which was opposed in principle to state industry and spared the Gobelins as an artistic rather than a commercial foundation.

The carpet loom was similar to the haute lisse *for tapestry, but the manner of weaving was entirely different. It employed peculiar carpet stitches. The weaver faces the front of his work and follows (far more literally than in tapestries) the pattern painted on a sheet of paper fastened right before his eyes. The paper is blocked out in squares of ten lines of which the vertical ones are repeated at the same intervals by varying the color of every tenth thread in the warp.*

a
f
h
a
d
f
d
f
a
g
e
d
g
i
i
d
b
g
g

Vol. IX, Tapis de Turquie, Pl. I.

Plate 342 *Carpets II*

A weaver lays the warp on a warping frame attached to a wall of the shop.

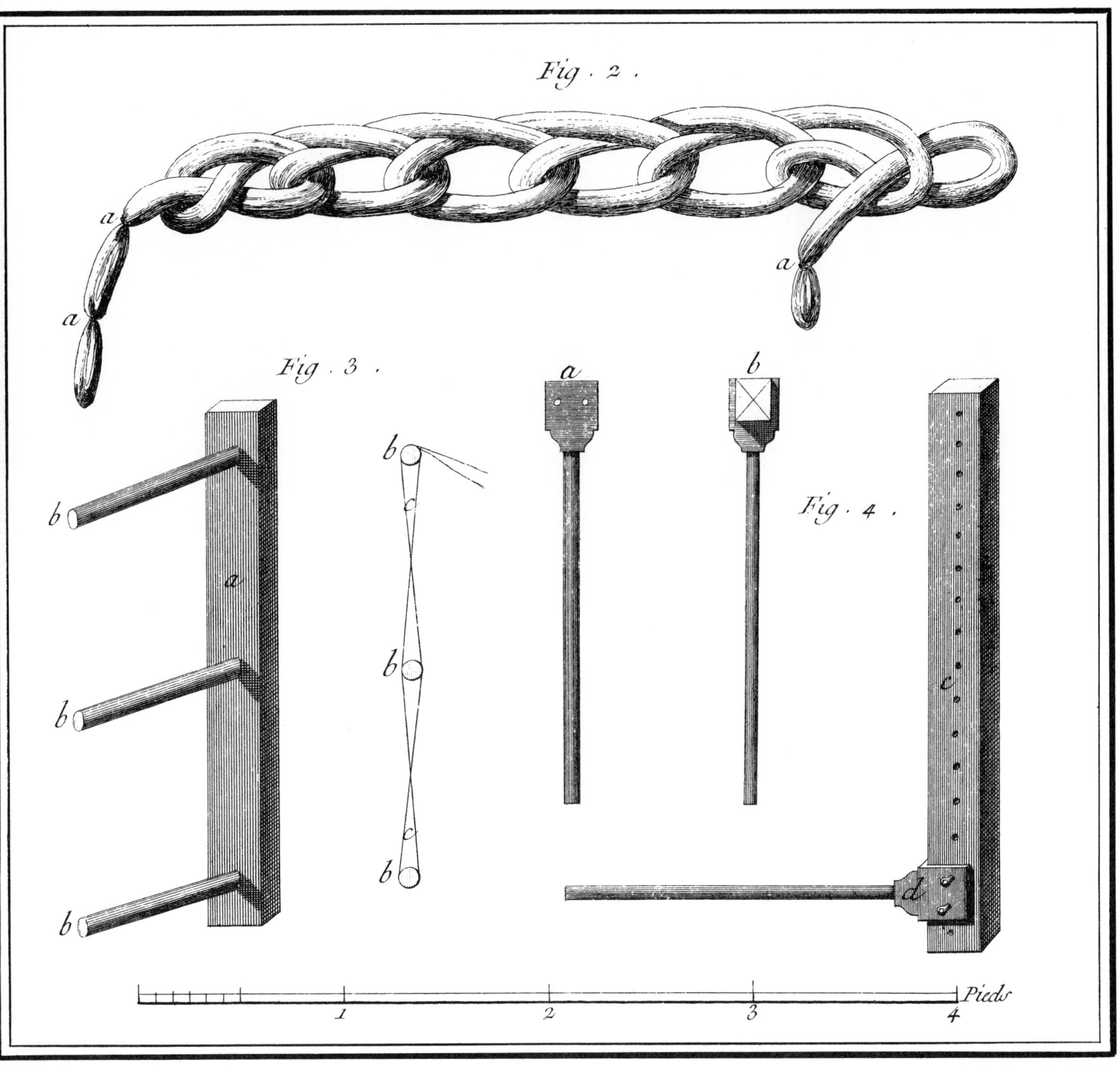

Vol. IX, Tapis de Turquie, Pl. IV.

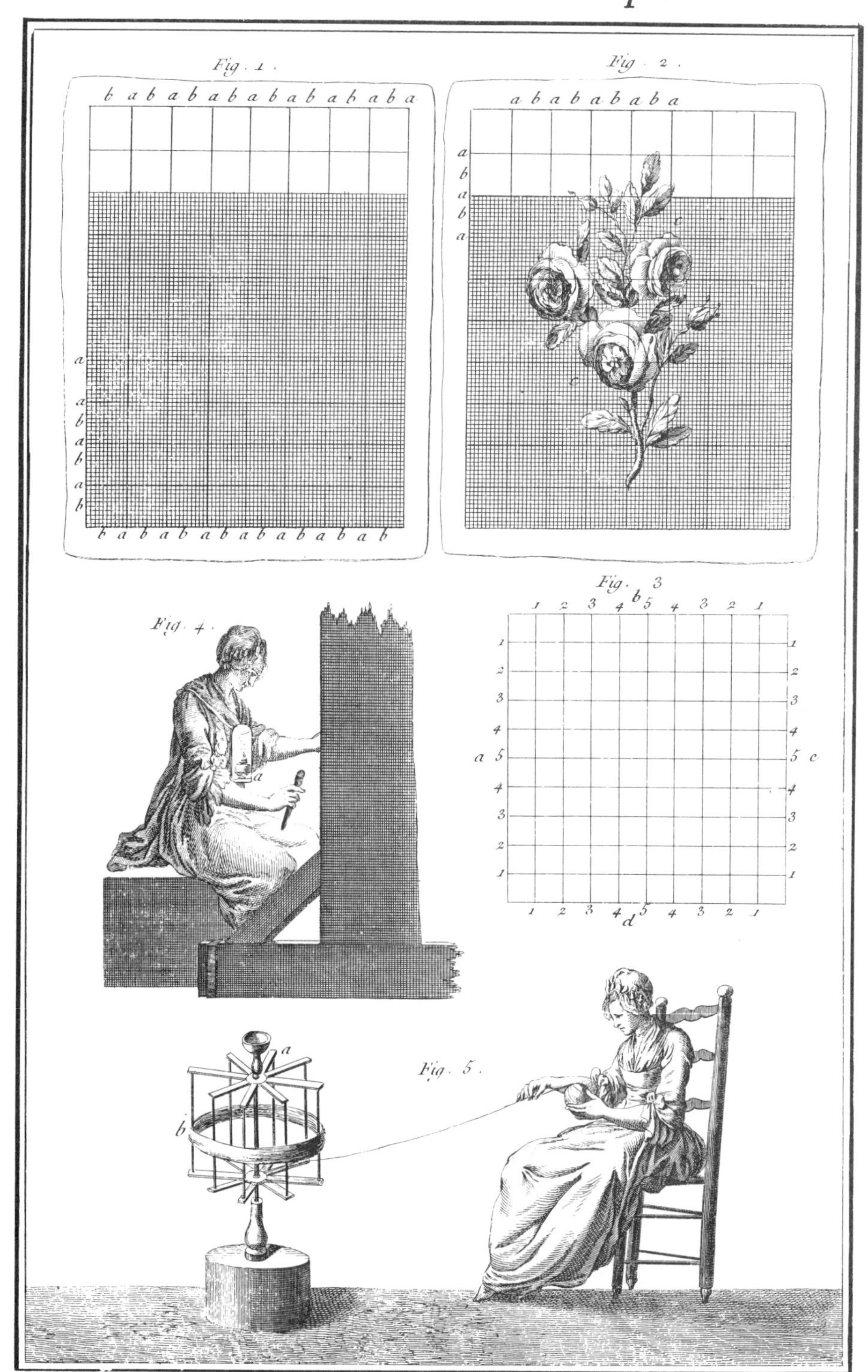

Vol. IX, Tapis de Turquie, Pl. V.

Plate 343 Carpets III

Figs. 1 & 2 show how a design of roses was traced on a piece of blocked paper to be repeated in the carpet. Each vertical line on the paper represents a thread in the warp. In Fig. 4 the female weaver works by the light of a lamp hung around her neck. Her colleague of Fig. 5 winds a skein of yarn.

Plate 344 Carpets IV

This illustration shows how a weaver made a knot and how he copied the paper pattern (Fig. 1, i). He is not, as might appear at first glance, a three-handed weaver. The operation to the right (Fig. 2) is separate. It consists of drawing out the velour thread for the braid.

Plate 345 Carpets V

Clipping thread ends (Fig. 1) and firming the weave with the comb (Fig. 2). The method of carpet-weaving, it is clear, results in a more solid texture and body in the fabric than would be needed or suitable in tapestries.

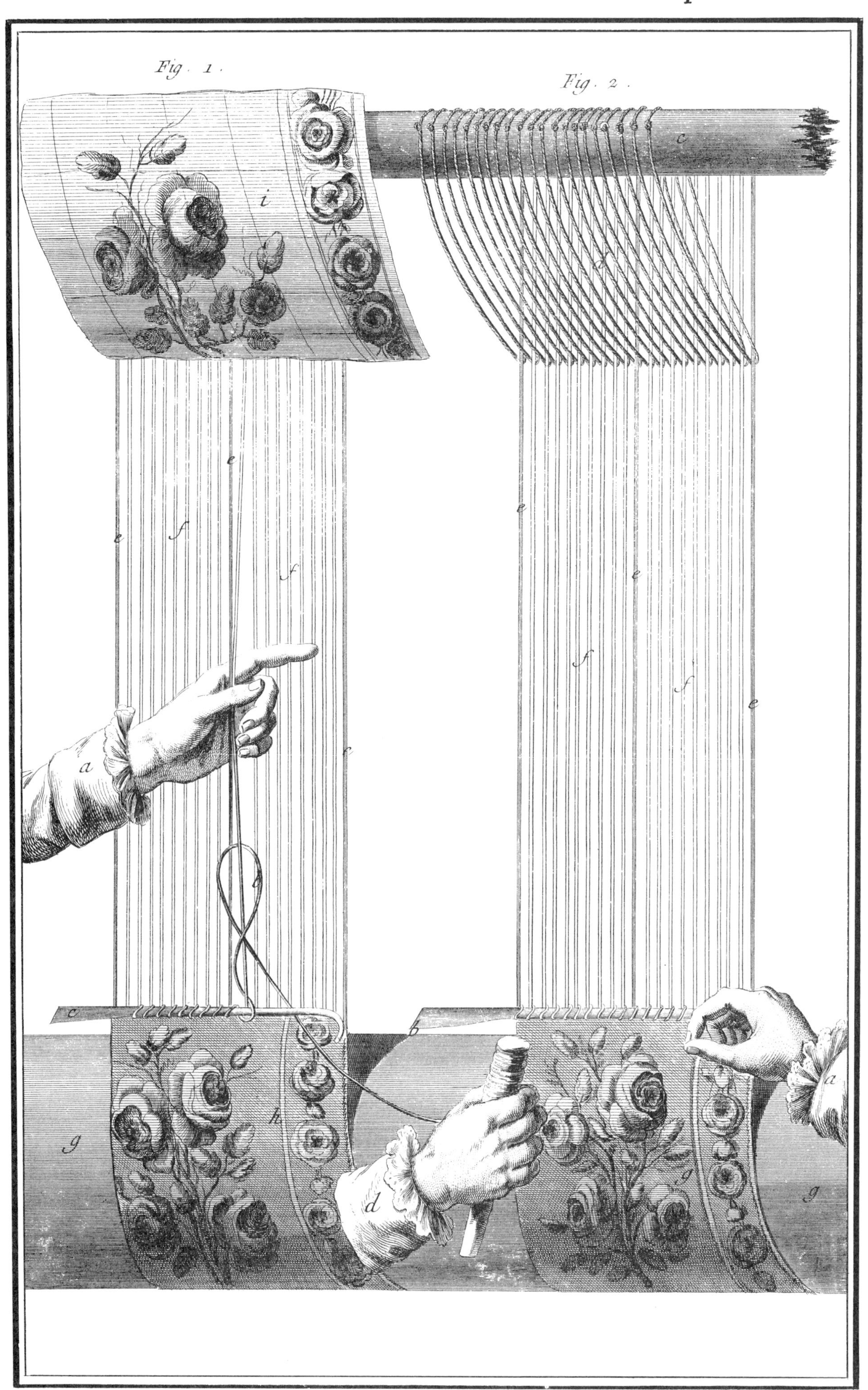

Vol. IX, Tapis de Turquie, Pl. VII.

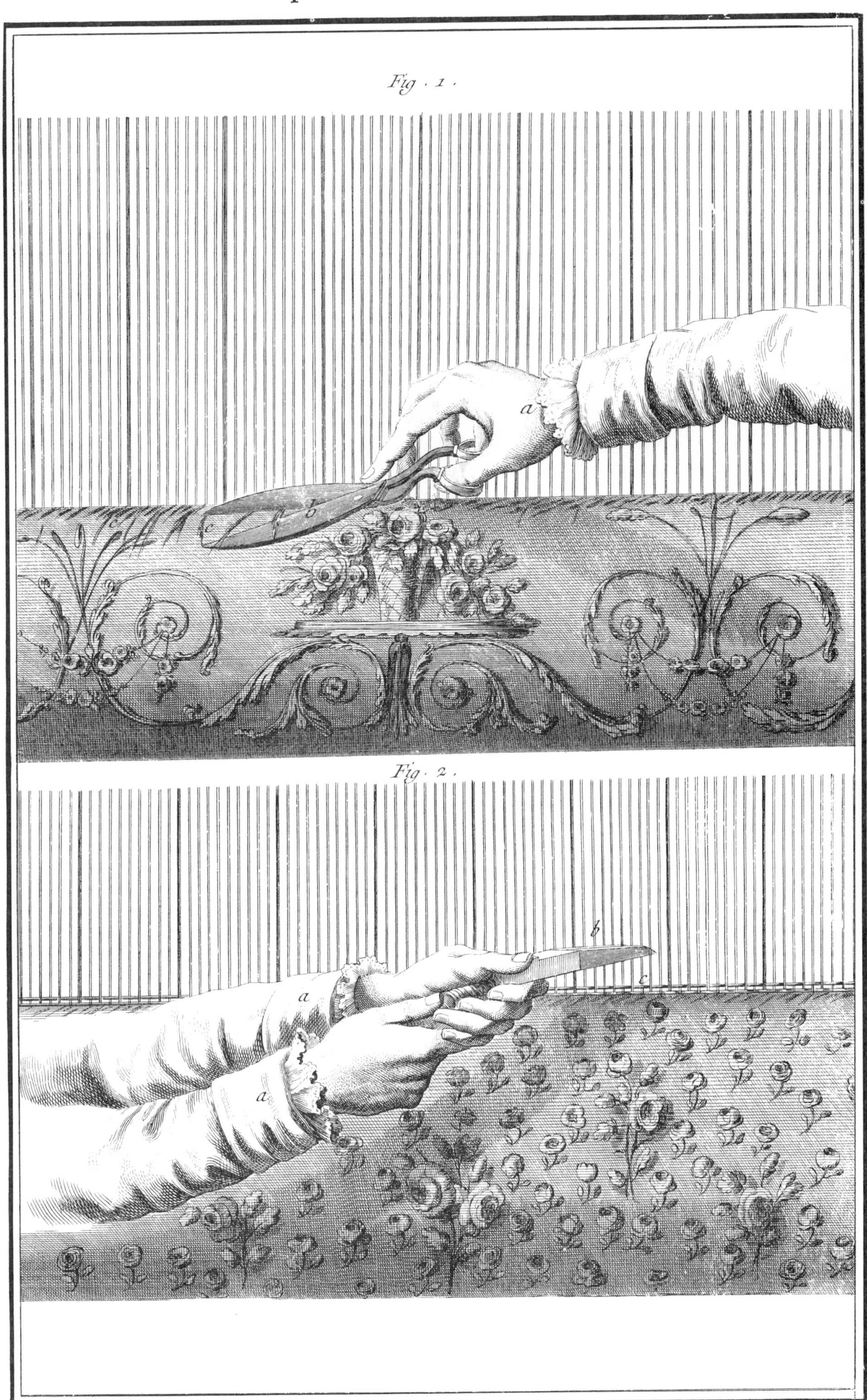

Vol. IX, Tapis de Turquie, Pl. VIII.

Plate 346 Upholstery I

Upholstery offered an important outlet for the weaver's art. Lacking inner springs, the 18th century had to rely on hair and woolstuffs for comfort. In this montage are displayed almost all that might be needed in the way of house-furnishings, all in the style of Louis XV.

There was in certain quarters a somewhat heavy-handed tendency to use tapestry motifs in upholstery. This rarely succeeded, since a design meant to confront the eye on some great wall across a majestic gallery would be rather less successful when confined to a chair cushion. The owner of this shop appears to have had a surer and more restrained taste, but he left certain chair frames uncovered (f) so that customers interested in display might order what they liked.

Plate 347 Upholstery II

Three of the many styles of bed canopies: the Polish style (Fig. 1); the Turkish (Fig. 2), which could serve by day as a divan; and the less exotic alcove preferred in France and England (Fig. 3).

Plate 348 Upholstery III

Upholstering a Louis XV armchair.

a
f
d
d
d
b
b
c
e
a
l
l
e
a
a
m

Upholstery I

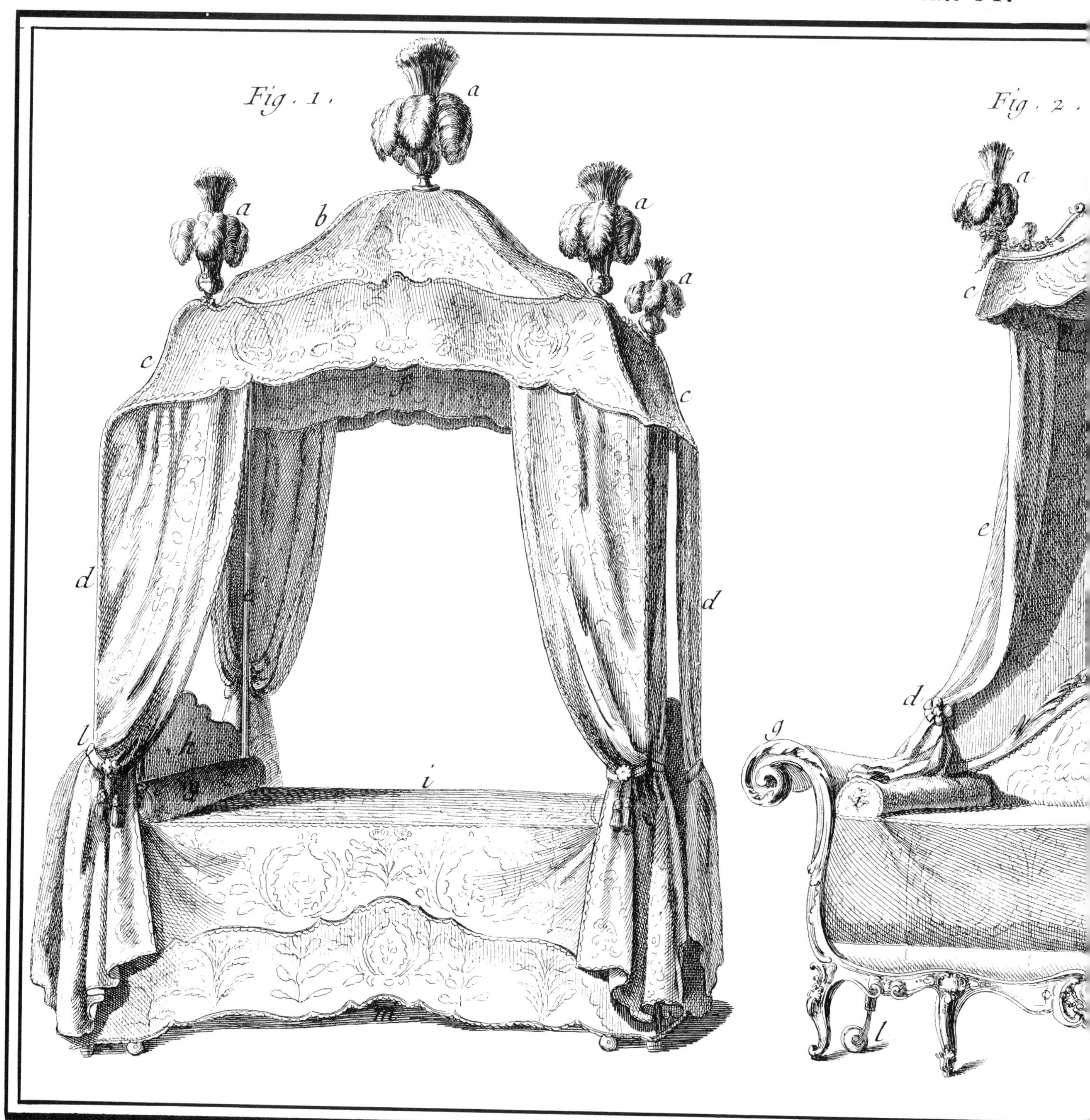
Fig. 1.
a
a
b
a
a
c
c
f
d
d
l
h
g
i
m
l
Fig. 2.
a
c
e
d
g
l

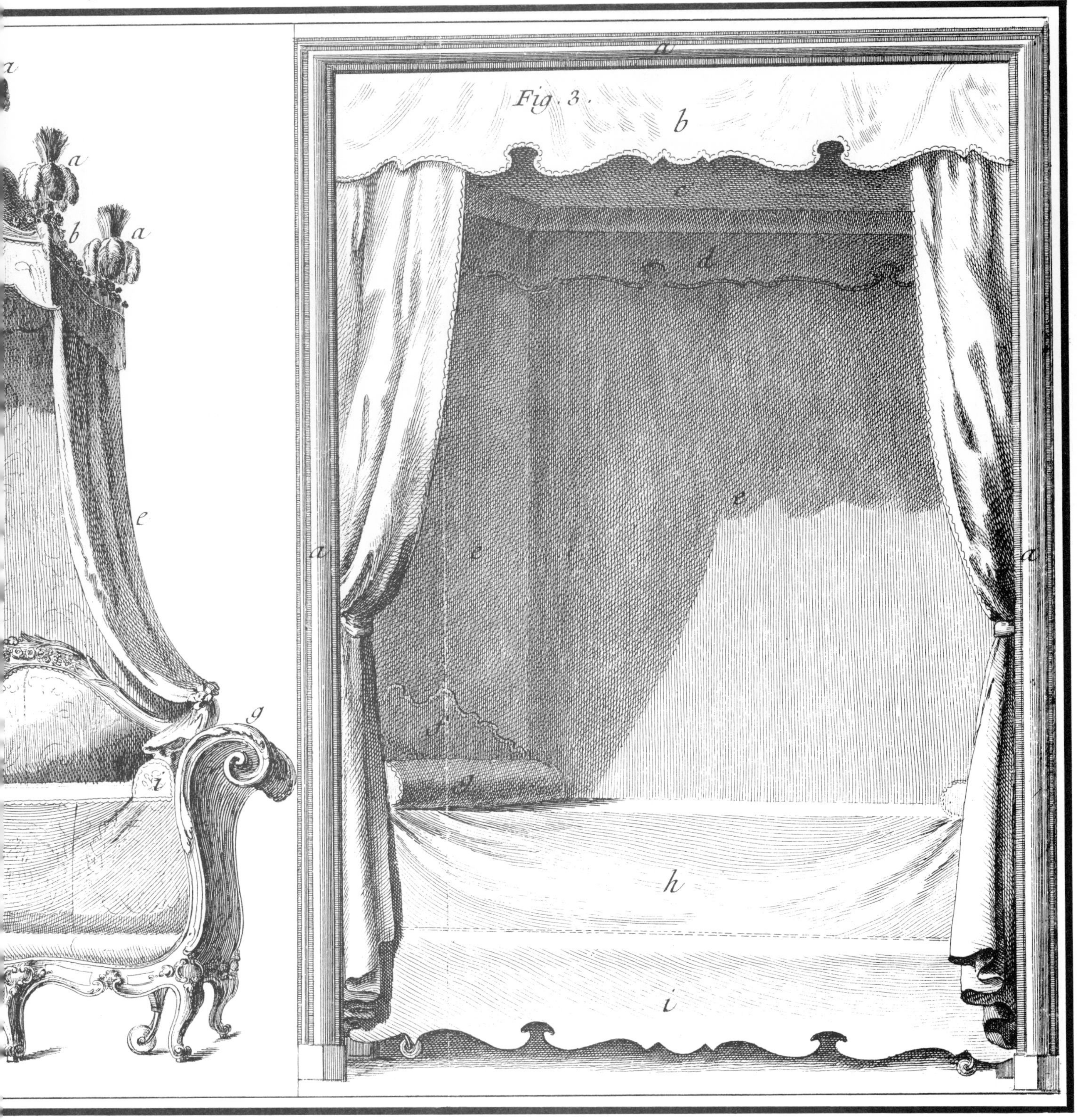

Vol. IX, Tapissier, Pl. V.

Fig. 1.
a
c
b
c
Fig. 2.
a
b
c
c
b
b
b
Fig. 3.
a
b
c
Fig. 4.
a
c
d
b
e

Plate 349 Dyeing I

Established in the buildings of the old Gobelins dyeworks, the Royal Manufactory carried on the business of its predecessors in conjunction with weaving tapestry. Dyeing was one of the trades for which Colbert's ministry established the strictest regulations. Among other things, it prescribed a sharp division between dyers of petite teinte licensed to work only in fugitive dyes on shoddy goods, and those who like the Gobelins practiced grande teinte and dyed sumptuous fabrics in fast colors. But dyeing is a trade which cannot be described connectedly, for its lore comprised a near infinity of detail. The very summary account in the Encyclopedia lists eighty-odd tints for wool, each produced by a different dye. There are for example, seven different dyestuffs for the nuances of red alone—each dye had to be processed differently depending on its particular properties. Not only so, but the fibers of wool, linen, cotton, and silk were affected very differently by the same dye stuffs, so that to duplicate in silk a certain shade of yellow in wool required another dye. The 18th century chemists wrestled manfully with this chaotic subject, but they could not aspire beyond a rational classification to a true rationalization of its principles.

All that can be illustrated, therefore, is the equipment and the main types of operations. The plate shows the workshop of the Gobelins giving on to the Bièvre through the door (g). Bolts of woolen cloth are washed in the cisterns in the foreground, from which the water is circulated through a system of hoses and valves (b, c) to a furnace which heats it for the dyeing vats (o). The water supply is replenished when needed by a series of reservoir tanks, one of which (i) projects into the shop.

Plate 349

Vol. X, Teinture des Gobelins, Pl. I.

Vol. X, Teinturier, Pl. VII.

Plate 350 Dyeing II

We are shown the method of preparing one dye, yellow safflower or "bastard saffron." It is a simple matter to extract the color in warm water. The dyestuff is pulverized (a) and then stirred about in tanks for which purpose the workers put on boots and tread it, supporting their own weight on ropes since the footing is a mucky one.

Plate 351 Dyeing III

The power plant which circulates water from the cisterns through the system of reservoirs and dye vats is in the room next to that on the first plate. The horses move a chain pump. The chain (g) and the liftpipe (h) rise at the right to the receiving reservoir (k).

n
n
n
n
d
c
e
c
m
b
a
a

Vol. X, Teinture des Gobelins, Pl. II.

Plate 352 Dyeing IV

The dye vat is equipped with a reel and crank (Fig. 1) for turning the goods evenly through the solution of steaming dye, but the workers also push and prod the cloth to make sure that folds and wrinkles do not prevent it from taking dye evenly. On days when the water of the Bièvre (Fig. 2) ran clean, workers might wash goods in the running water.

Vol. X, Teinture des Gobelins, Pl. VIII.

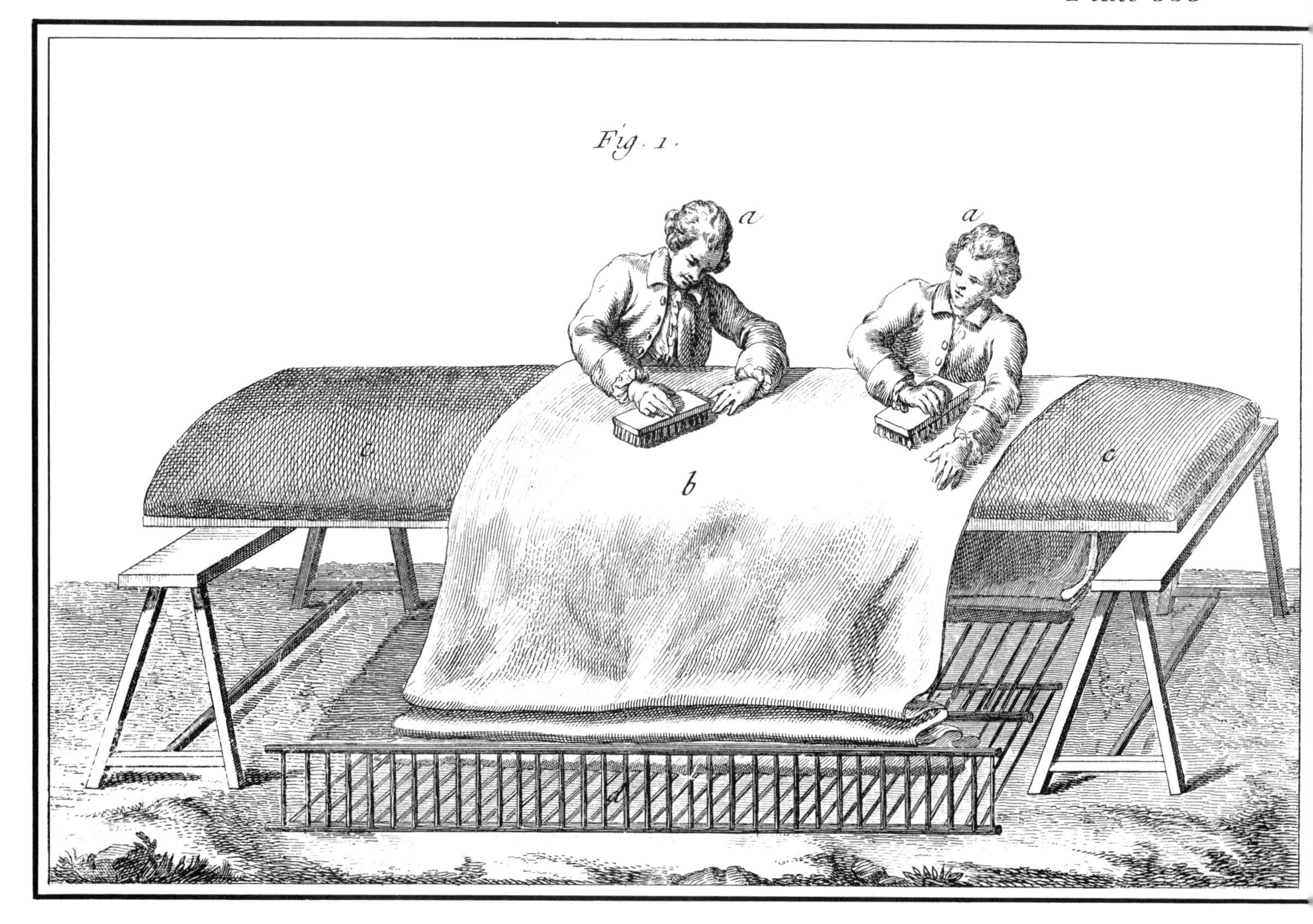

Plate 353 Dyeing V

After being dyed, the surface of woolen goods would look matted and encrusted with bits of dye and foreign matter. The nap is raised by brushing with a stiff brush (Fig. 1). Opposite, (Fig. 2) are drying racks.

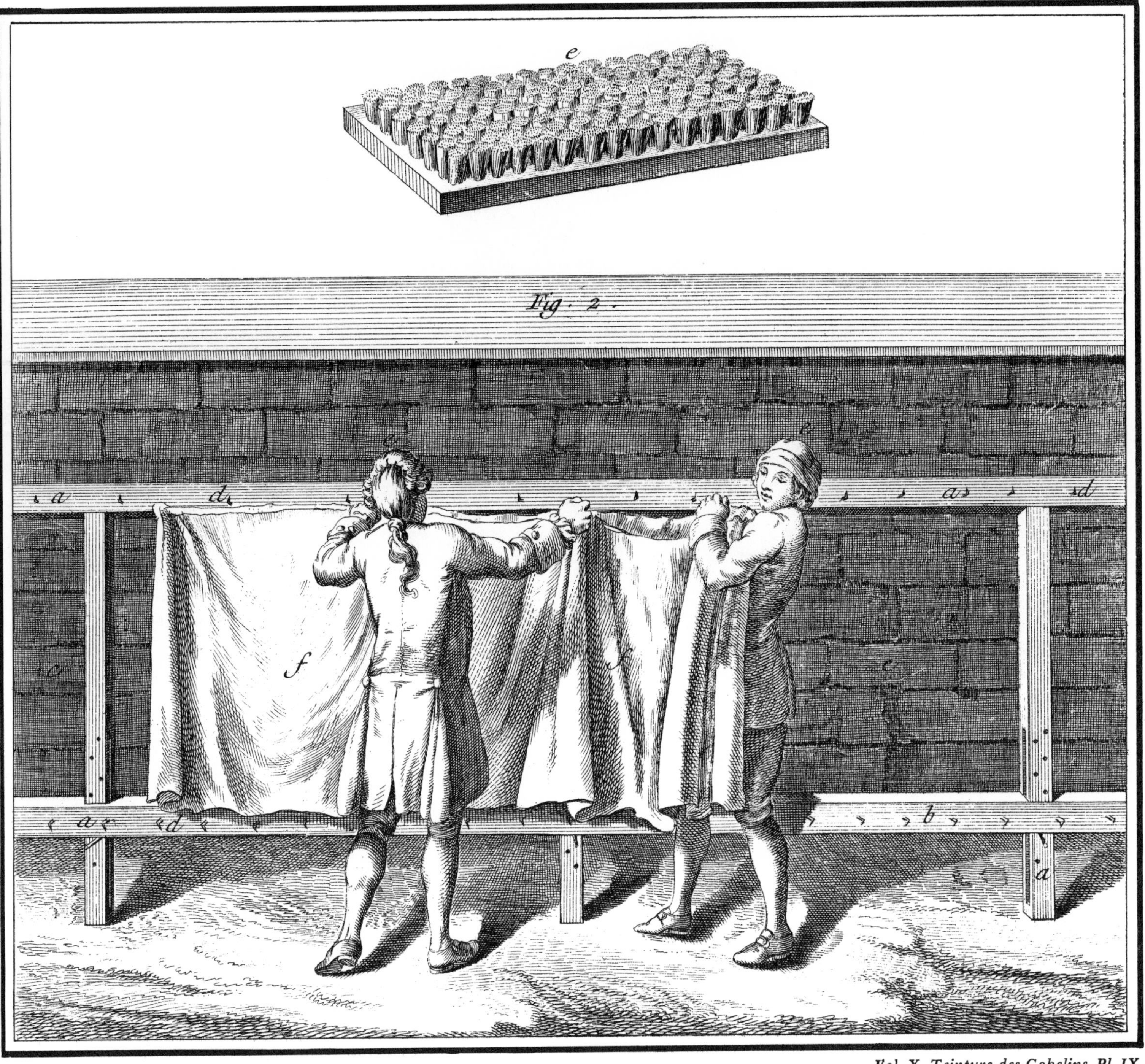

Vol. X, Teinture des Gobelins, Pl. IX.

Vol. X, Teinture des Gobelins, Pl. X.

Plate 354 Dyeing VI

In the manufacture of tapestries, it was the yarn that was dyed, not the finished cloth. The exact color was absolutely essential, and after the hanks were dyed (c) a checker would go down the rack with a color chart (d) to verify each one.

Vol. X, Teinture des Gobelins, Pl. XI.

Plate 355 Dyeing VII

The drying room is heated by a coal stove in all seasons. When the hanks of dyed yarn are dried, they are twisted (d) into tight rolls for delivery to the weaving shops where they will await distribution amongst the tapestry weavers. That each twist contains a full measure of yarn is verified on the scale before it goes to the weaver, who would be seriously embarrassed should a skein give out unexpectedly. It is obvious from their dress that these are a class of employees altogether inferior to that to which the weavers belonged by virtue of skills and training.

Plate 356 Dyeing VIII

Silk required a very different treatment from wool, and dyers tended to specialize in one material or the other. This is a silk-dyer's, with its washtubs of copper (i), well-bucket to dip water from the river below (p), furnace and boiler (a, c), and dye vats equipped with running hot water (b).

m
s
h
g
v
i
i
i
i
t
l

Vol. X, Teinturier de Riviere, Pl. I.

Plate 357 Dyeing IX

In order to preserve the lustre of silk, which was, of course, its distinctive quality, the skeins of thread had to be boiled in soapy water for a period of three hours before dyeing—until the silk which was floated by its film of dirt and grease when first put into the cauldron "sinks like lead." Then it was ready to take a dye and hold it.

To prevent the skeins from twisting and tangling during the boiling, they were packed in cloth bags (d), one of which (g) is being fished off the bottom. Once boiled, the skeins were ranged on wooden rods (c) to be passed through the dye vat (b).

fig. 1
a
b
c
d
e
f
g
h
i

Plate 358 Dyeing X

The silk-dyer moors a barge in the river for his workers. We are not told what river, though it would probably be the Seine. In any case it is not the little Bièvre, and we are no longer at the Gobelins.

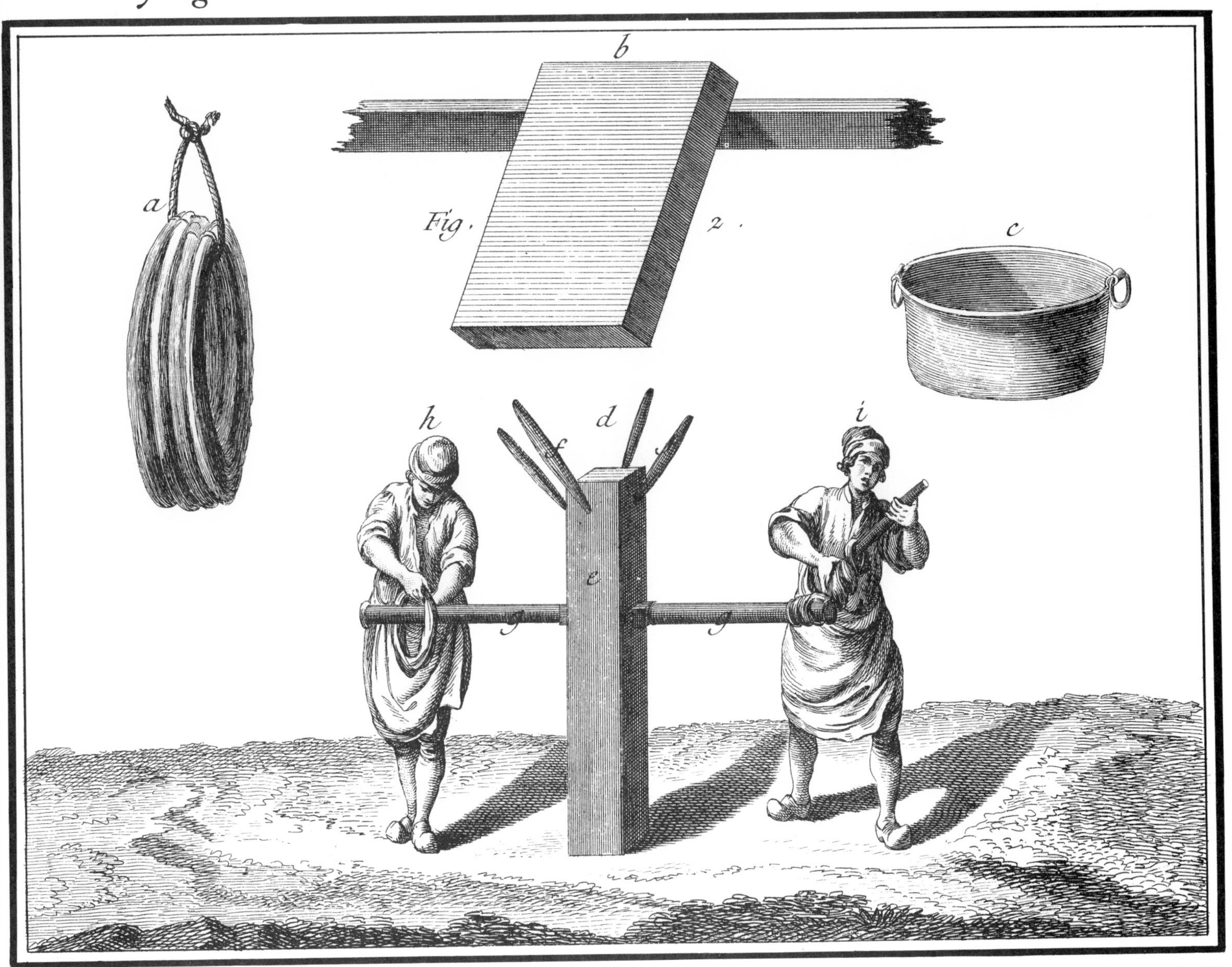

Vol. X, Teinturier, Pl. VI.

Paper & Printing

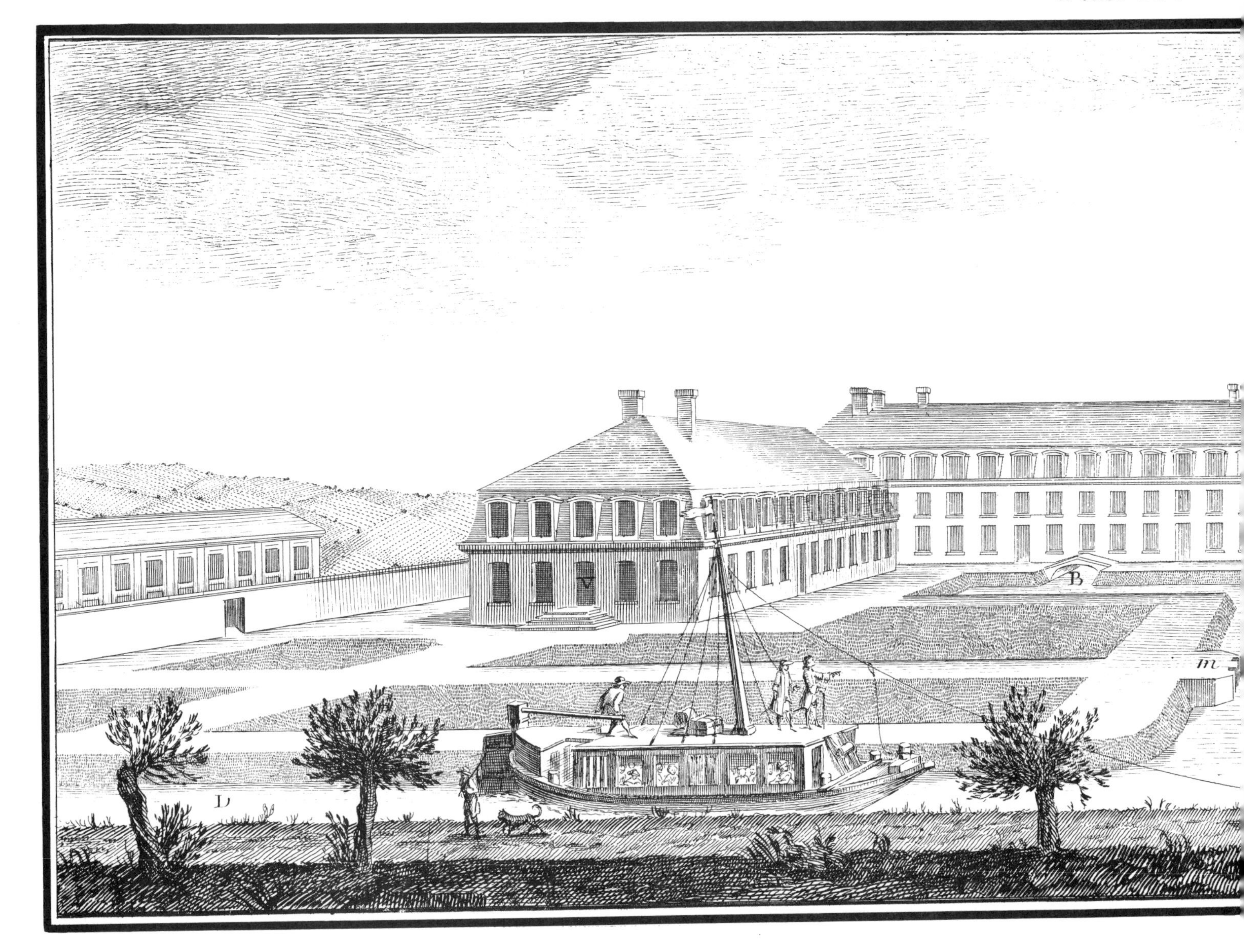

Plate 359 Paper I

About 60 miles south of Paris, near the city of Montargis, stood the great paper factory of l'Anglée, an impressive example of large-scale industry.

France was the greatest producer of printing paper in the 18th century. Although some connoisseurs found the writing paper of Holland to be of finer grade, others held that this superiority was a myth. In any case, a number of French manufacturers successfully deceived the public by using watermarks bearing the legend Papier de Hollande *and it is not now necessary, perhaps, to settle the dispute, the less so since Holland and France still produce the finest handmade papers.*

Papermaking requires unlimited supplies of water, here obtained from the river Loing at Montargis.

This stream formed part of the navigation system of the first of the great canals, the Briare, which linked the Seine and the Loire (see Plate 479). A branch is led right into the factory. One arm (B) served the department where cellulose fibres were prepared, and the other (G) that which molded the sheets. From time to time it was necessary to drain the basin, which could be closed by a watergate (m, m). Like many paperworks, this was a highly paternalistic establishment. It provided housing right on the job for its workers in the outbuildings on either flank.

The canal boat appears to be carrying passengers rather than freight—Parisian tourists, perhaps, or commercial travellers, who are having the main features of the factory explained to them.

On the next page is a view of one of the water wheels that power the machinery of the plant, portrayed from the downstream side.

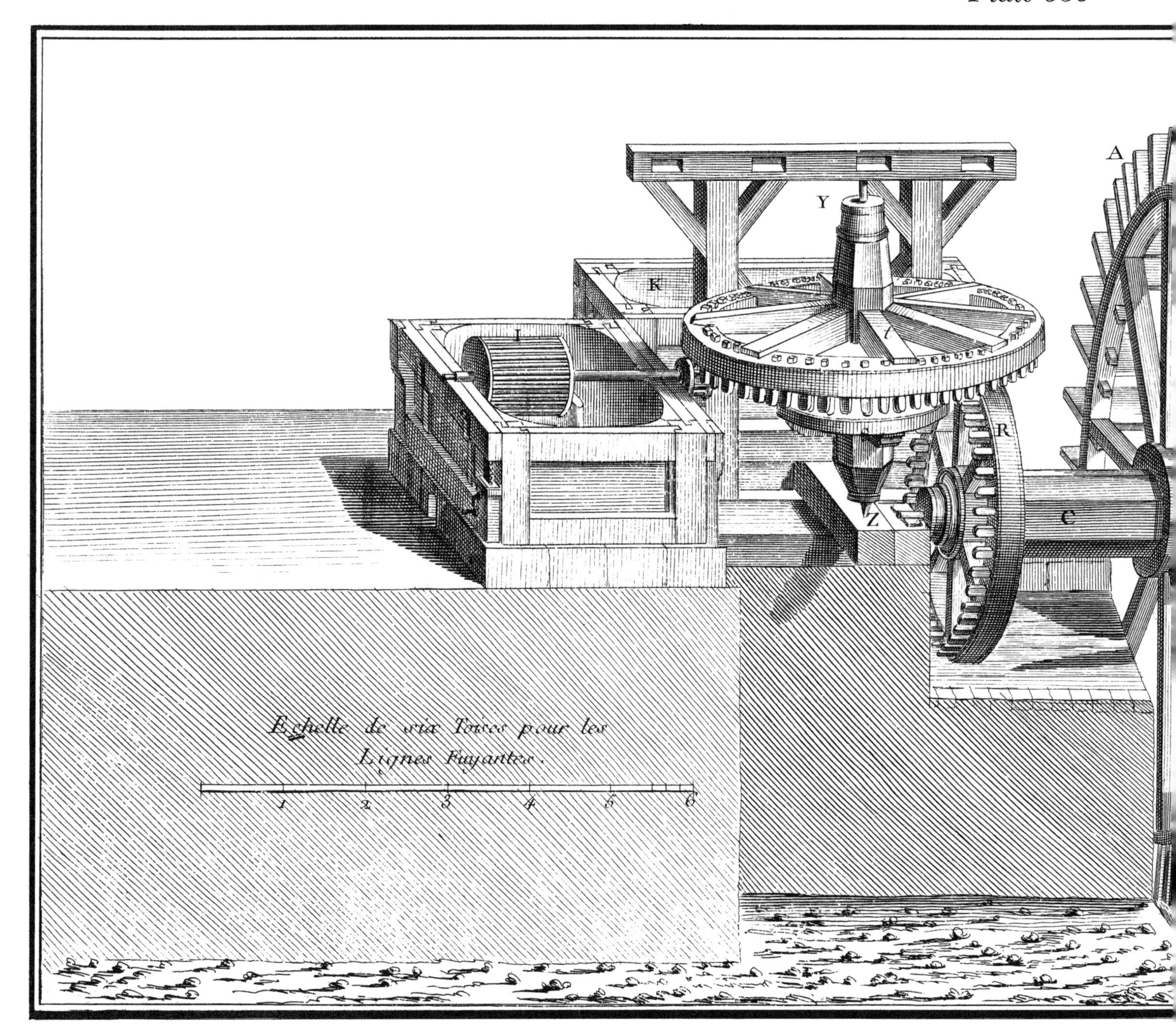
A
Y
K
I
R
Z
C
Echelle de six Toises pour les
Lignes Fuyantes.
1
2
3
4
5
6

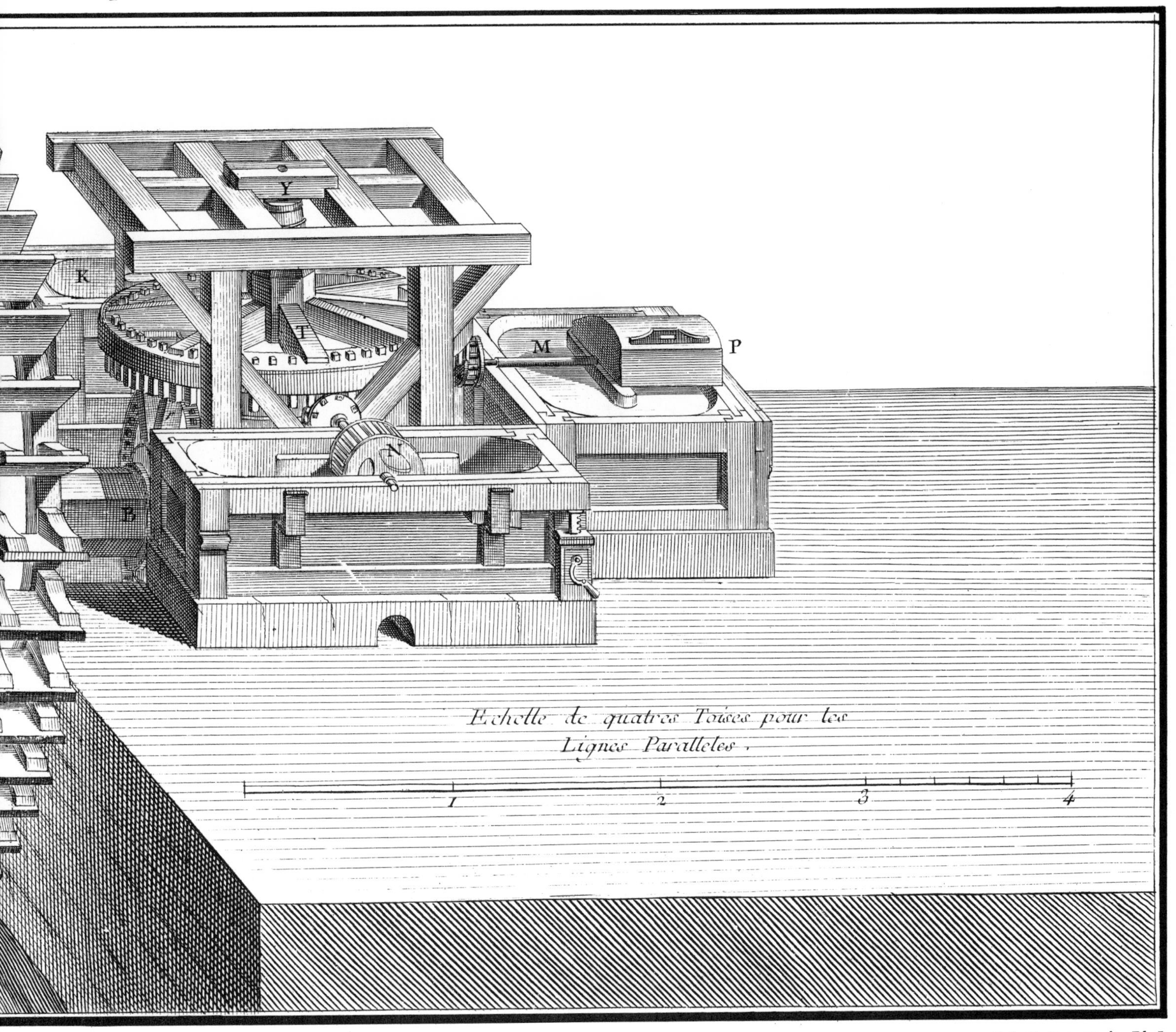
Y
K
T
M
P
N
B
Echelle de quatres Toises pour les
Lignes Paralleles.
1
2
3
4

Plate 360 Paper II

The origins of paper are lost in the mists of Chinese antiquity, whence it passed to the Arabs in the 8th century—as a result it is said, of the successful siege of Samarkand by the forces of Islam in 751 A.D. Paper was introduced by the Moors into Spain in the 12th century, and its manufacture established in Italy by the late 13th, again thanks to the Arabs, but by way of Sicily rather than Spain. It was in general use in Europe for literary purposes by the 14th century, replacing skin as the material on which to write.

All paper is made of fibers of cellulose drawn ultimately from the woody portions of plants. Its manufacture requires two basic stages: first, the fibers must be separated mechanically and so treated as to eliminate all the accompanying non-cellulose matter; secondly, the pulped fibers are sieved from a water suspension in a thin sheet and matted on a mold of wire mesh.

In the 18th century rags of linen or hempen cloth were still the source of cellulose. "RAGS make paper, PAPER makes money, MONEY makes banks, BANKS make loans, LOANS make beggars, BEGGARS make rags,"—so went an old saying.[1] The advantage of rags was that the process of making textiles had already eliminated from the fibers the extraneous matter of which the modern manufacturer must rid his wood-pulp.

The significance of paper for the diffusion of culture needs no laboring. But it may be that the substitution of wood pulp for rags in the 19th century will prove even more revolutionary than the late medieval displacement of parchment by paper. For it is the latter innovation which alone permits mass production of reading matter. Whatever may prove true culturally of modern literature, it is literally true that much of our writing will not last.

The vignette represents the ragpickers' shop—they are sorting out these raw materials from pile D into bins. Below appears a general plan of the factory of l'Anglée, with the water moving in from the Montargis canal in the channel A and the flow dividing to the two millraces. We are now at the left hand side of the plant, in the area MMMM. It is apparent from the scale that the main body of the factory is about 450 feet in length, and each of the wings about 150 or slightly less. The grounds are planted in parterres. This is almost an industrial chateau.

1 *Quoted by Dard Hunter,* Papermaking *(New York, 1947), an excellent and fascinating history.*

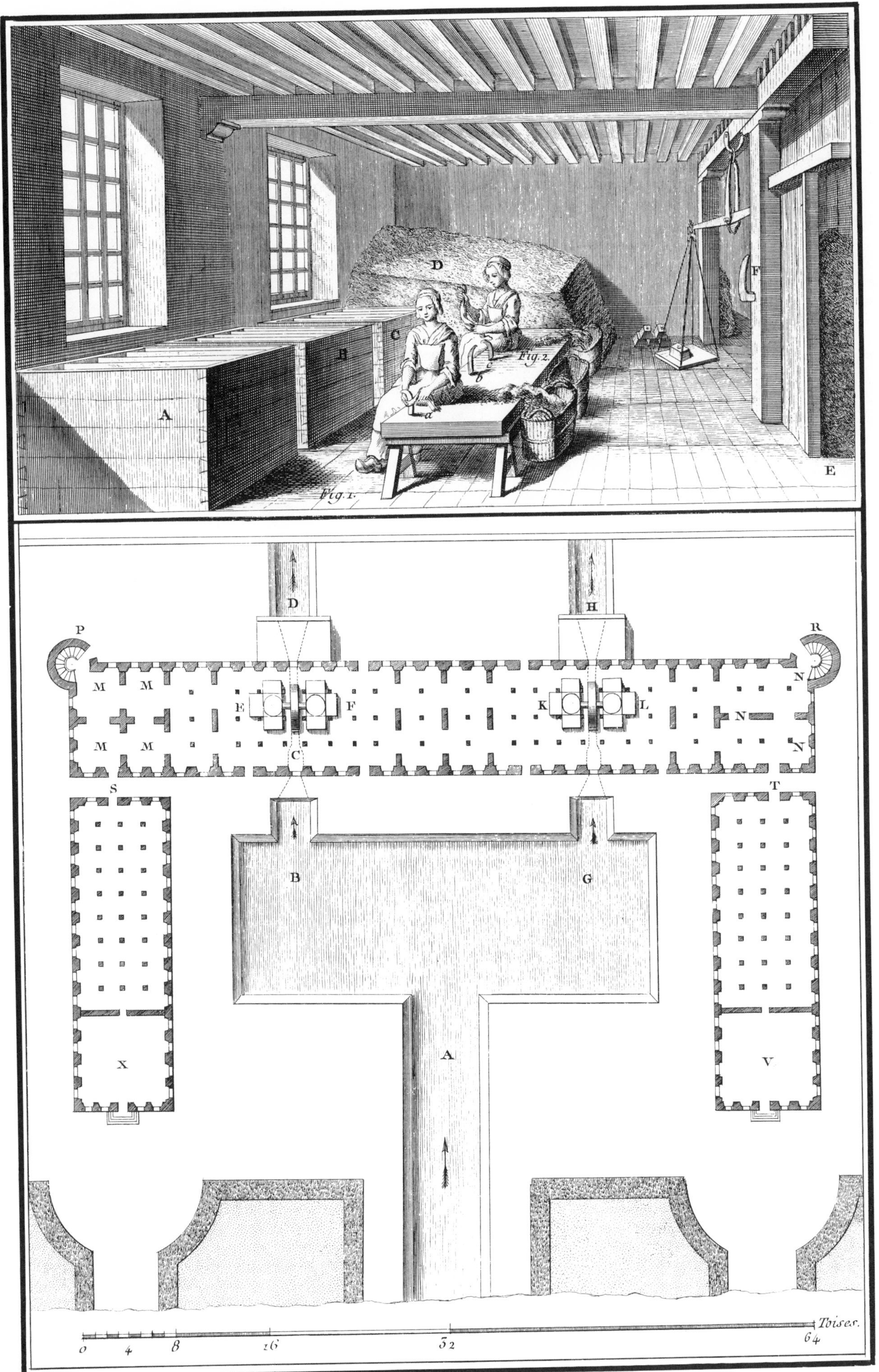
D
C
B
A
Fig. 2.
a
b
F
E
Fig. 1.
D
H
P
R
M
M
M
M
E
F
C
K
L
N
N
N
S
T
B
G
A
X
V
Toises.
0
4
8
16
32
64

Plate 361 Paper III

Rags disintegrate more readily if they are rotted beforehand for six weeks or two months. That is the purpose of this shop, where a moist binful (K) of dejected rags is fermenting sourly. In some countries lime was added to hasten the process, but the result was weakened fibers and inferior paper, and in France, always seeking the middle way between progress and quality, the use of lime was forbidden by government regulation. After two months or so, the mass of rags will have taken on a bilious yellow look, which in the final product is transformed pleasingly into the slightly ivoried appearance of good 18th century paper.

The part of the plate opposite gives a bird's-eye view of the stamping-mills to which the rags will next be passed. A small stream of water (1, 2, 3) is diverted from the millrace (DFE) to aliment the troughs or "vat-holes" (M, M), the mortars in which the stampers (Figs. 8-11) pound up and down like mechanized pestles and macerate the rags.

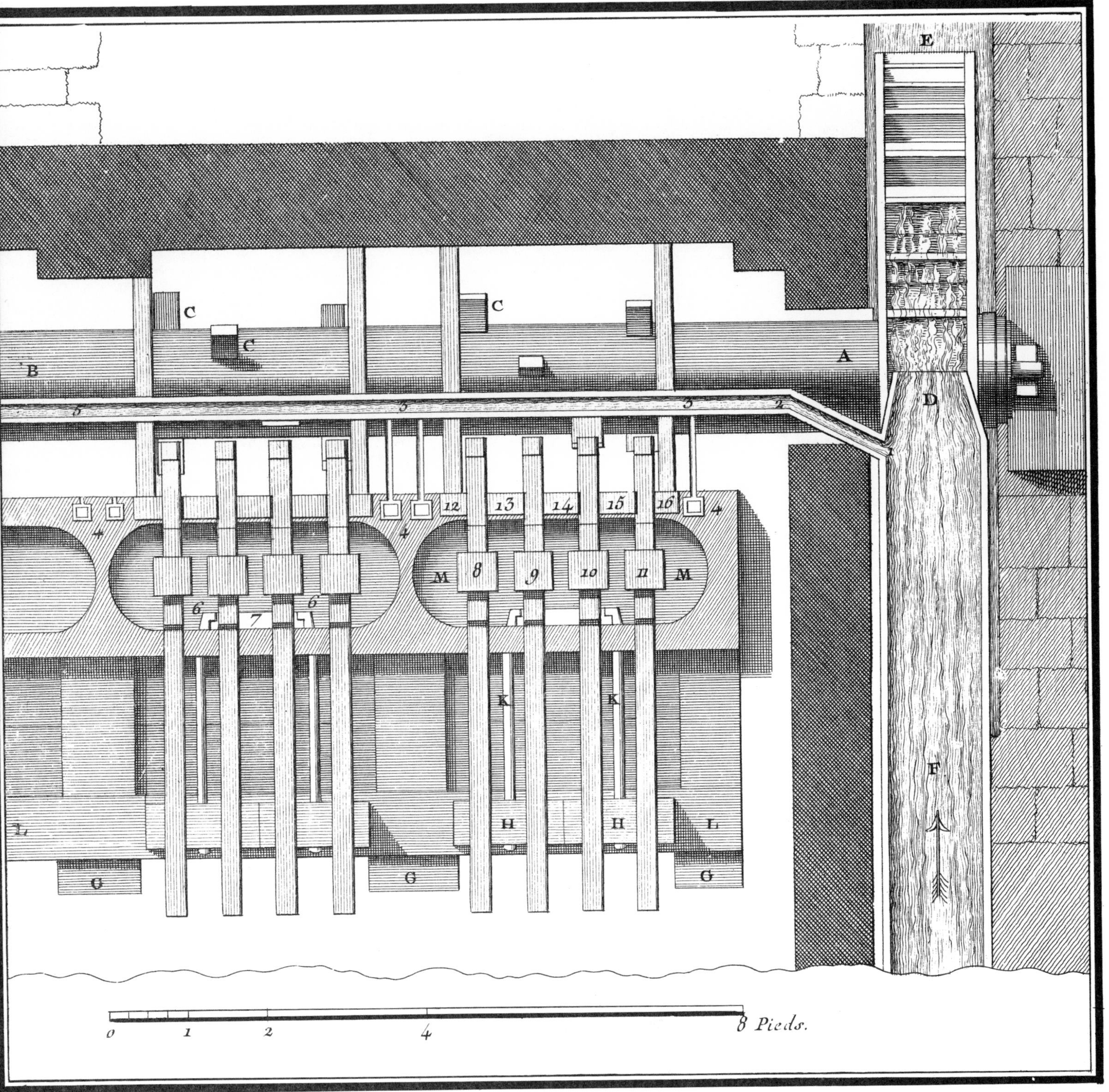

Vol. V, Papetterie, Pl. II.

Plate 362 Paper IV

In the l'Anglée mill one further step intervened between fermentation and maceration. Before the flaccid, rancid rags are thrown into the mill, they are cut up manually in a shop manned by boys. Since boys differ in size, the manufacturer is solemnly advised to have on hand blocks on which the smaller boys can stand up to their work (Fig. 3).

It is not very difficult work. All the boys do is hack at the rags with a scythe-blade, which must be sharpened frequently (Fig. 4, f) because of the dulling effect of wet rags. The bottom of each bin is a sort of grill, through which bits of dirt and other foreign bodies are supposed to fall out.

Opposite we are given another view of the stamping mill, this time in profile.

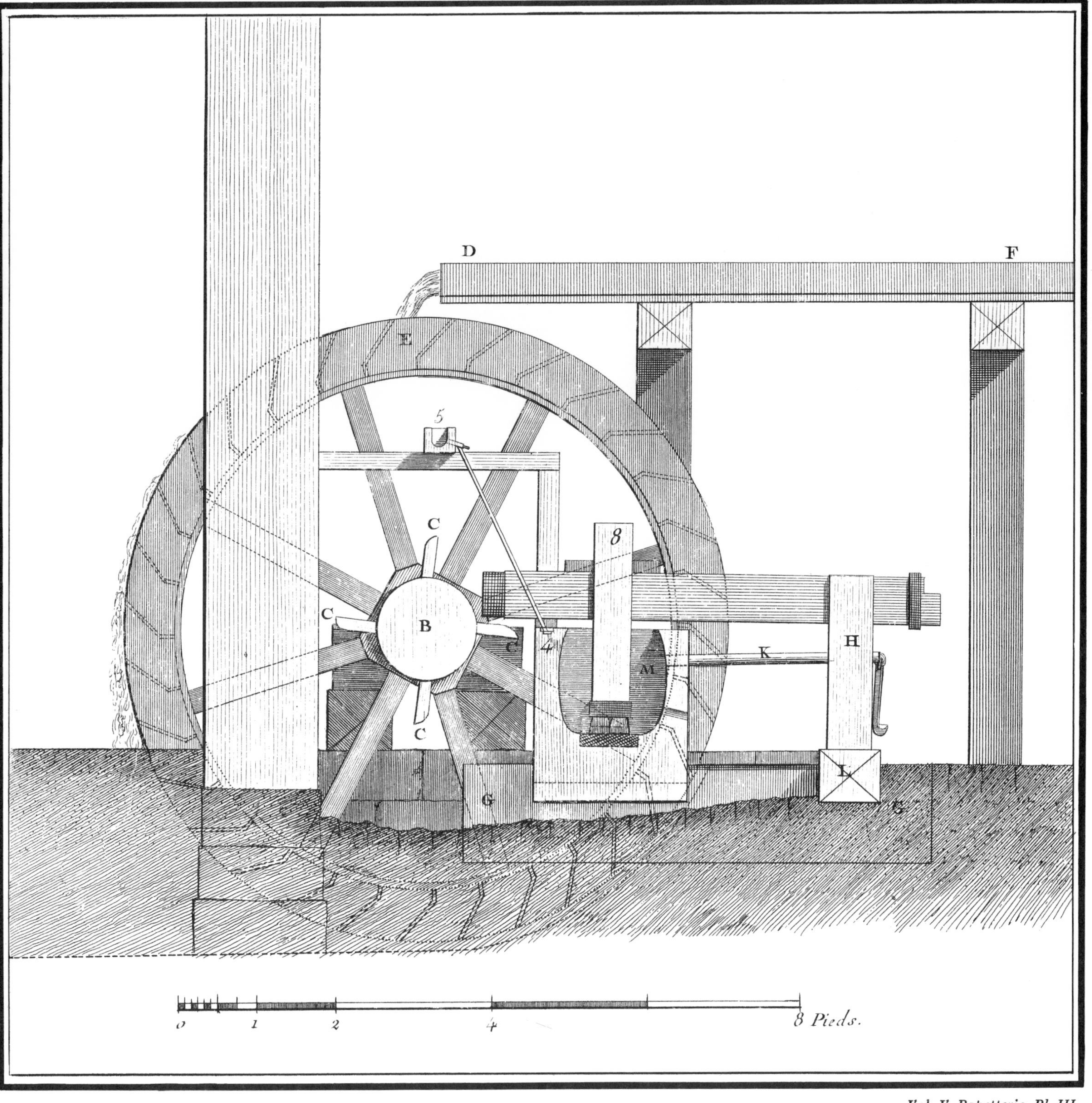
D
F
E
5
C
8
C
B
C
4
K
H
M
C
G
L
0
1
2
4
8 Pieds.

Plate 363 Paper V

The stamping-mill which beats the fibers to a pulp—the origin of that expression—consists of three sets of four pestles each. The ends of the far two are tipped with spikes which further tear the rags apart, and the troughs into which they pound are irrigated with water led in (Figs. 1-3) from the millrace (Figs. F, D). This washes away impurities. The mass of "half-stuff," as it is called by the time it reaches the second set of stampers, is then reduced to free fibers of cellulose in the last, where the stampers are blunt and where no wash water is used for fear of washing the pulp itself away.

Vol. V, Papetterie, Pl. IV.

Plate 364 Paper VI

Papermaking, in a sense, simply transformed woven textiles into very thin felts which were formed by passing a wire mold through a bath of pulped fibers and drying out the sheets so picked up.

The manufacturing end of the plant occupied the right hand apartments of the l'Anglée establishment (Pl. 360, K, L, NNN). This is the shop that makes the molds, which are frames of wire mesh laid across ribs of heavier gauge. The frame itself was oak. It was of the greatest importance that the mold be skillfully made, for it had to stand constant immersion in baths of hot water without warping. It was used, moreover, with a rim called a "deckle edge" which served as the sides of a scoop and made the edge of each sheet. These deckles had to fit exactly and to be interchangeable from one mold to another.

This type of mold was in general use until about 1800. If paper of the 18th century is held to the light it is possible to see in it the outline of the ribs and the finer cross-mesh of the "chain line." Experts can often identify the date and maker from these patterns, even when the manufacturer did not use a distinguishing watermark, as the l'Anglée plant apparently did not. This is too bad, for the plates do not show how the pattern of the watermark would be woven into the mold in wire thread.

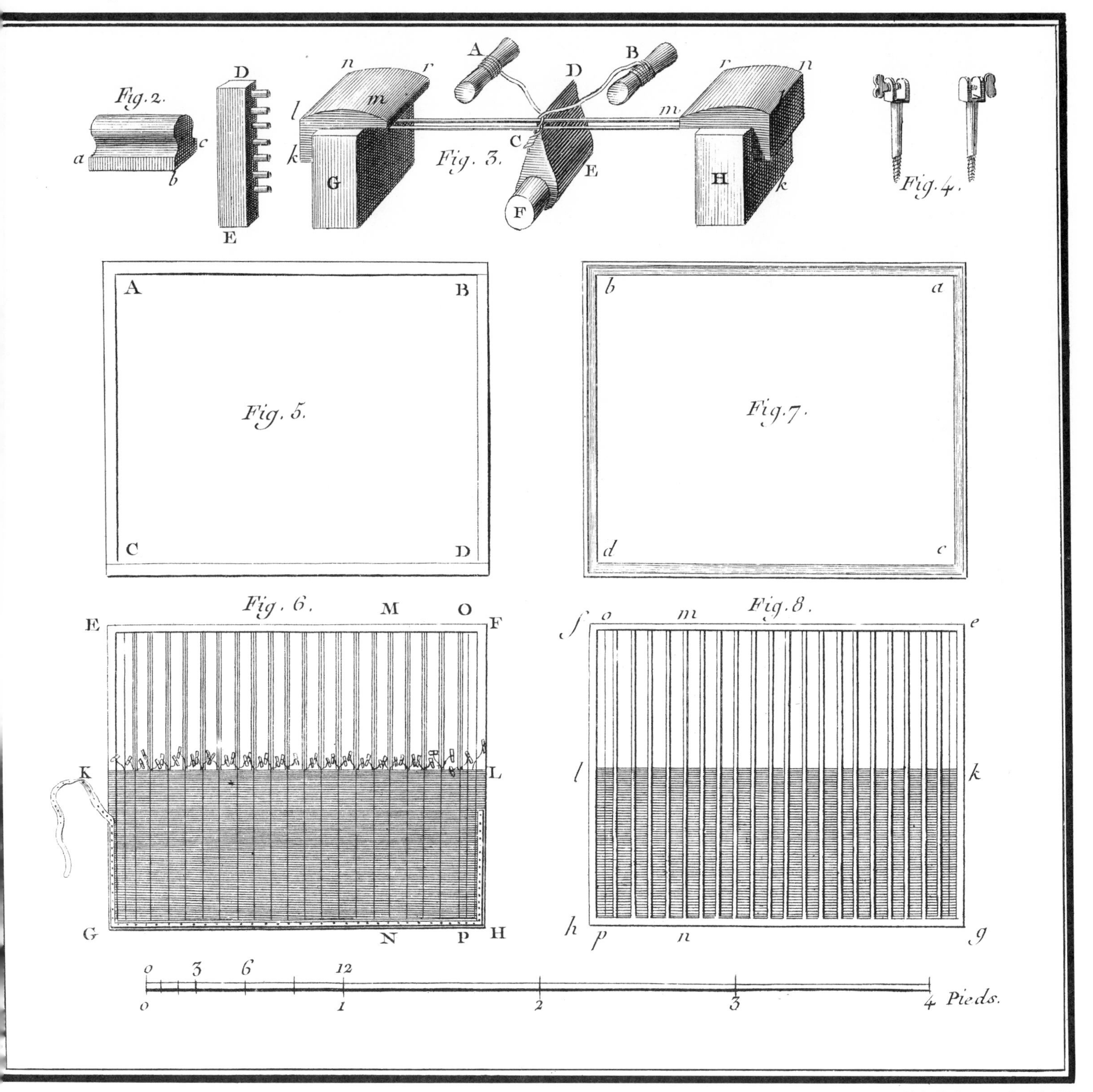

Vol. V, Papetterie, Pl. IX.

Contrary to common belief watermarks did not originate as trade-marks, although that is what they have become. They were simply symbols, though of what is a much conjectured question.

Plate 365 Paper VII

Here is the heart of the papermill, where sheets of paper are dipped from a vat (Fig. 1) filled almost to the brim with the fibrous mush from the stamping-mill. A team of three men attend each vat. They had to work together in almost mechanical rhythm. The vatman set the pace. His was a skilled and arduous job. Sleeves pushed back, he takes a mold, plunges it steeply into the vat, turns it horizontally and lifts, sieving out a layer of pulp and letting water drain back into the vat. With almost the same motion he gives his mold two shakes—from right to left and from front to back. This matted the fibers both ways and gave the sheet equal tensile strength.

Next the vatman slides his mold along the plank to his partner (Fig. 2), who couches the sheet. The coucher flips the mold over onto a piece of felt slightly larger than the paper. Then he covers the wet sheet with another felt, takes a second sheet from the vatman, and in this way builds up a blotting pile of alternating felt and paper until he has accumulated 260 sheets.

260 sheets made ten quires which, in the French measure, comprise a "post" or half-ream, to be squeezed in the press of Fig. 4. Unmaking the post afterwards and separating the felts and sheets was the "layer's" job (Fig. 3). After he has accumulated eight posts, they are again pressed, this time without the felts. Detail of the equipment ap-

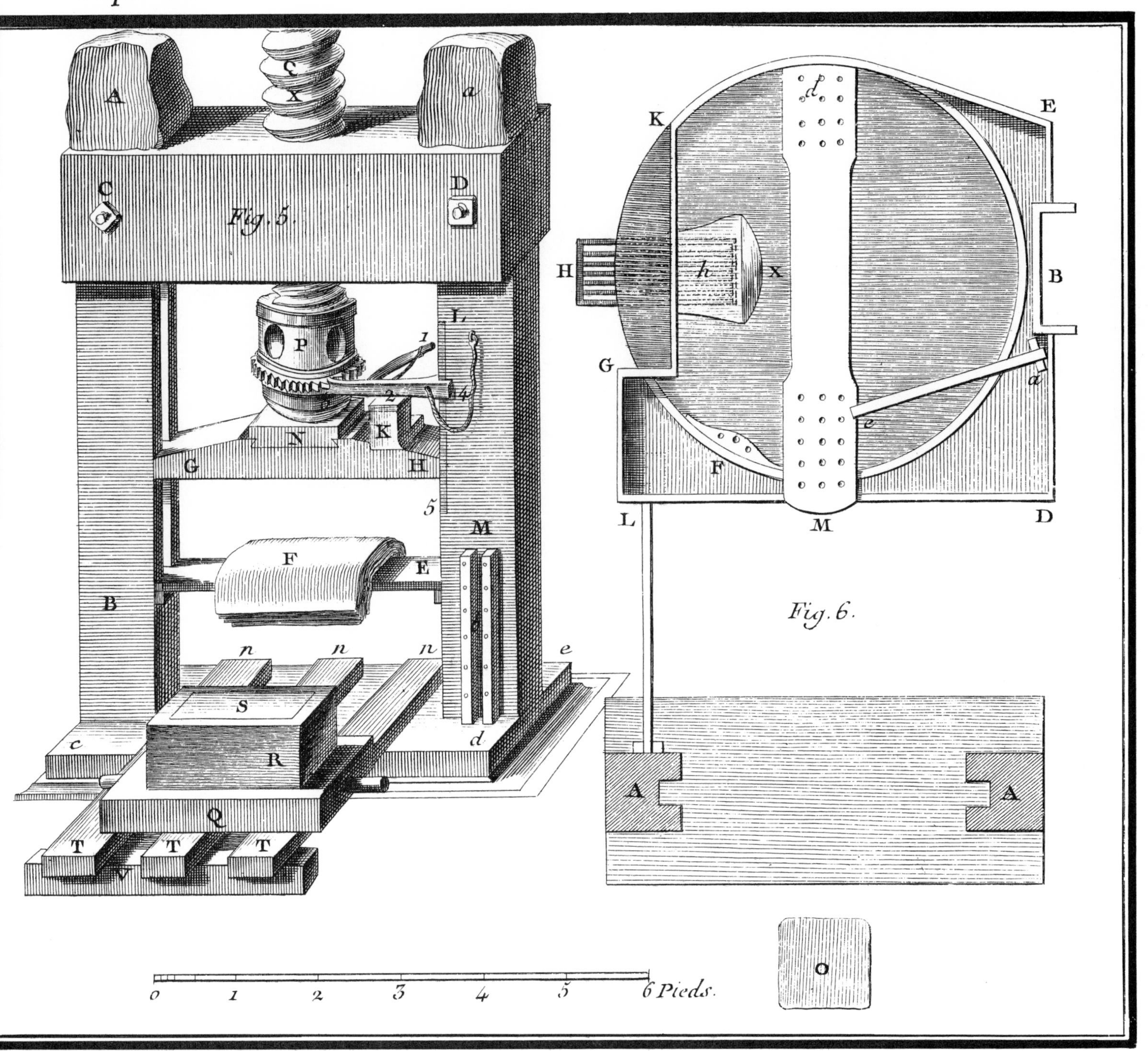

Vol. V, Papetterie, Pl. X.

pears above. Fig. 6 shows the vat from the top. In a day's work of 16 hours a good team could turn out 16 posts or eight reams a day, which amounted to 4,160 sheets of paper.

Vatmen were liable to a peculiar form of paralysis, an occupational hazard which was the consequence of the debilitating atmosphere of humidity and heat and of the monotony of the four exhausting motions of their job—dip, lift, shake right and left, shake to and fro, 4,160 times a day. Sometimes they would recover after a layoff, but often they would have to be put to other work, for theirs was a selective paralysis which forbade them only this particular sequence of motions.

Vol. V, Papetterie, Pl. XI.

Plate 366 Paper VIII

In order to take ink without running, paper had to be sized with a gelatinous animal glue. Paper manufacturers made their own sizing from waste scraps of hides which they would buy up from tanneries.

The basket (E) was loaded with hides, lowered into the cauldron (K) and boiled, and the sizing filtered (Fig. 1) through woolen cloth into a great copper storage tank. All vessels in this shop were of copper.

The sizing kettle itself (Fig. 2) had to be kept moderately warm by a small brazier. The sizer adds a little alum and copperas before dipping the paper. He handles a "post" at a time, holding the paper with two batons in his left hand while he riffles the pages

Vol. V, Papetterie, Pl. XII.

with the right. He suspends the paper a few minutes in the bath, changes his grip to the other end, withdraws the post from the tub, and lays it in the press beside him. When he has sized 12 posts, he turns to the press (Fig. 3) and squeezes out the excess sizing. But this must be a gentle compression, lest all the sizing be extruded.

Plate 367 Paper IX

After sizing, the paper is dried sheet by sheet, an operation requiring space, care, and patience, but not much skill.

Plate 368 Paper X

By now we have travelled right across the main building of the factory into the hall of the right-hand wing, which contains the finishing room. Here sheets are inspected (Fig. 1), polished (Fig. 2), folded (this for some reason is a little girl's job, Fig. 3), counted out into bundles of 25 folios (Fig. 4), and powerfully compressed in order to crease and package in reams of 20 bundles or 500 folios.

The machine which rendered handmade paper obsolete, except for certain very special purposes, was itself invented by a Frenchman, Nicholas-Louis Robert, who took out a patent in 1798. Its essential feature was that it substituted for the hand-mold a wire-mesh endless belt travelling across the vat. Pulp was poured onto the mesh by means of little buckets fitted onto a revolving cylinder which dipped them down into the vat and brought them over the belt. The belt itself supplied the "shake" (though only in the direction of motion, for which reason it is easier to tear machine-made paper one way than the other), and then carried the sheet under a felt roller.

The innovation consisted, not in a new principle of papermaking, but in mechanizing the old process. Since all processes are governed by the nature of materials, something of the sort is true of almost all the inventions which moved particular industries into the machine age. Many writers and critics of industrialism find it impossible to resist a temptation to sentimentalize old methods, and to discern rich human qualities in handicrafts. But what led Robert to seek a method of making paper by machinery was his disgust with the bad temper, quarrelsome obstinacy, and chronic drunkenness of the vatmen, couchers, and layers of the old papermakers' guild. He had no thought of cheapening the price of paper, much less its quality.

Indeed, the substitution of wood pulp for rags—an absolute necessity, of course, for mass production—was what coarsened and weakened modern paper, and not the replacement of man by machine. On the contrary, it is the opinion of Dard Hunter, whose historical Papermaking *(New York, 1947) is both charming and authoritative, that modern machine-made rag papers are in many ways superior to the handmade papers of olden times. They can be made perfectly square, of uniform thickness, and (best of all) even on the edges. It was an unavoidable imperfection of handmade paper that the rim or "deckle" of the mold left a rough edge to the sheet. This deckle edge was always trimmed off the best papers, and it is an instructive and ironical example of the snobbery of the antique that in modern times machines have actually been devised to counterfeit this defect, just as they have been to fake the "bull's-eye" which marred handmade "crown" glass (see Plate 235).*

Vol. V, Papetterie, Pl. XIII.

Plate 369 Printing I

The 18th century was an age of elegant bookmaking and skillful printing. Indeed, whatever pains the publisher of the present book may take, it is not likely that he will achieve so fine a production as the Encyclopedia *itself. It may, therefore, be consoling to reflect that, because of the excellence of 18th century paper and binding, the original text and plates will probably be in good condition centuries hence when these reproductions will have crumbled to dust.*

Printing, too, was at a very high level, much higher, indeed, than the printing of the following century. Type faces were graceful, well-designed, and pleasing to the eye. Pages were admirably planned, perfectly squared, and clearly pressed.

The first step in preparing type for printing was manufacturing punches for individual letters of a font. The punch was made of hard, fine steel, a bar of which is being finished at the forge (Fig. 1). This is to be a letter "B", and in Fig. 2 the artisan stamps what will be the loops of the "B" into the heat-softened punch-head, using a counterpunch. A glance at the facing part of the plate will clarify these steps. Fig. 5 is the vise in which the punch is held for stamping, Fig. 1 is the counterpunch, and Fig. 2 (below) shows the punch as it leaves the hands of the stamper (Fig. 2 above). Finally, in Fig. 3, above, the outer edges are delicately filed to leave the "B" as it appears below in Figs. 3 and 4.

This required great precision, and a perfect alignment of the axis of the letter. The filer guides himself by means of the device shown in Fig. 7, covering the surface of the stone (Fig. 8) with a film of olive oil to lubricate the filing. Then, he tests the punch by smoking its end and making an impression on a sheet of paper. Finally, the punch must be annealed to give it the hardest possible temper.

Unfortunately, there is no plate to illustrate the punching of the matrix. The operation was a simple one, however. The letter was struck into a blank of copper approximately an inch long by a quarter-of-an-inch wide and a sixteenth-of-an-inch thick. This matrix, carrying the negative of the letter, then slipped into the head of a mold assembly in which the letters were finally cast in type metal.

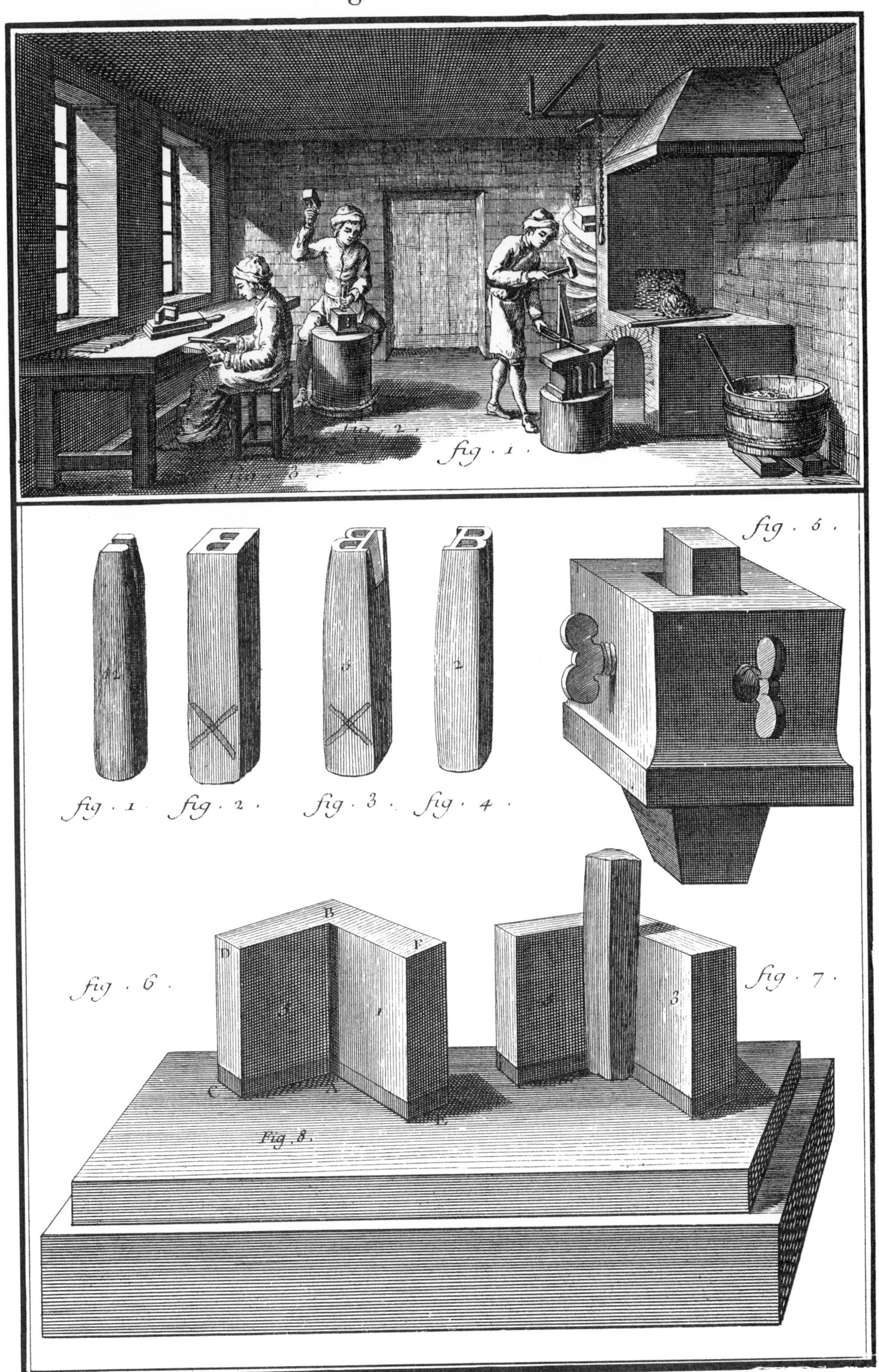
fig. 1.
fig. 2.
fig. 3.
fig. 1.
fig. 2.
fig. 3.
fig. 4.
fig. 5.
fig. 6.
fig. 7.
Fig. 8.
B
D
F
C
A
E
3

Plates 370, 371, 372 Printing II, III & IV

Type metal is a low-melting alloy made by adding molten antimony (Fig. 3) to lead containing a little tin (Fig. 4). It is poured into ingot molds in front of the furnace, to be remelted as needed on the founder's stove at the left (Fig. 7). Three type-founders work around the stove. One (Fig. 5) dips a spoonful of molten type metal into his hand-mold. The second (Fig. 6) sets the metal into the matrix by a good, sharp shake; and the third (Fig. 8) is about to take out the letter. It requires only a moment for the type metal to harden. An experienced founder could turn out 400 letters an hour. On emerging from the mold, the letter carries a stem or "tang" which has to be broken off (Fig. 1). The girl in Fig. 2 then smooths the two sides of the letter by rubbing them against the surface of the circular stone. A few details of the mold (Pl. 371-2) may be of interest. Figs. 12 and 13 give a view of a matrix, with the letter punched into it. Fig. 1 shows the ensemble of the mold, with the matrix in place and its butt end just showing at (y)—the tab (4) for handling it is of sheepskin, and it is fastened by means of the padlock-like steel loop (ECD). The mold is filled with molten type metal through the square cavity at the top which gives onto the face of the matrix. As is apparent, the mold was a complex precision instrument—it consisted of over a dozen major parts and a large number of bolts and screws, each of which demanded very fine machining.

Vol. II, Fonderie en caracteres, Pl. I.

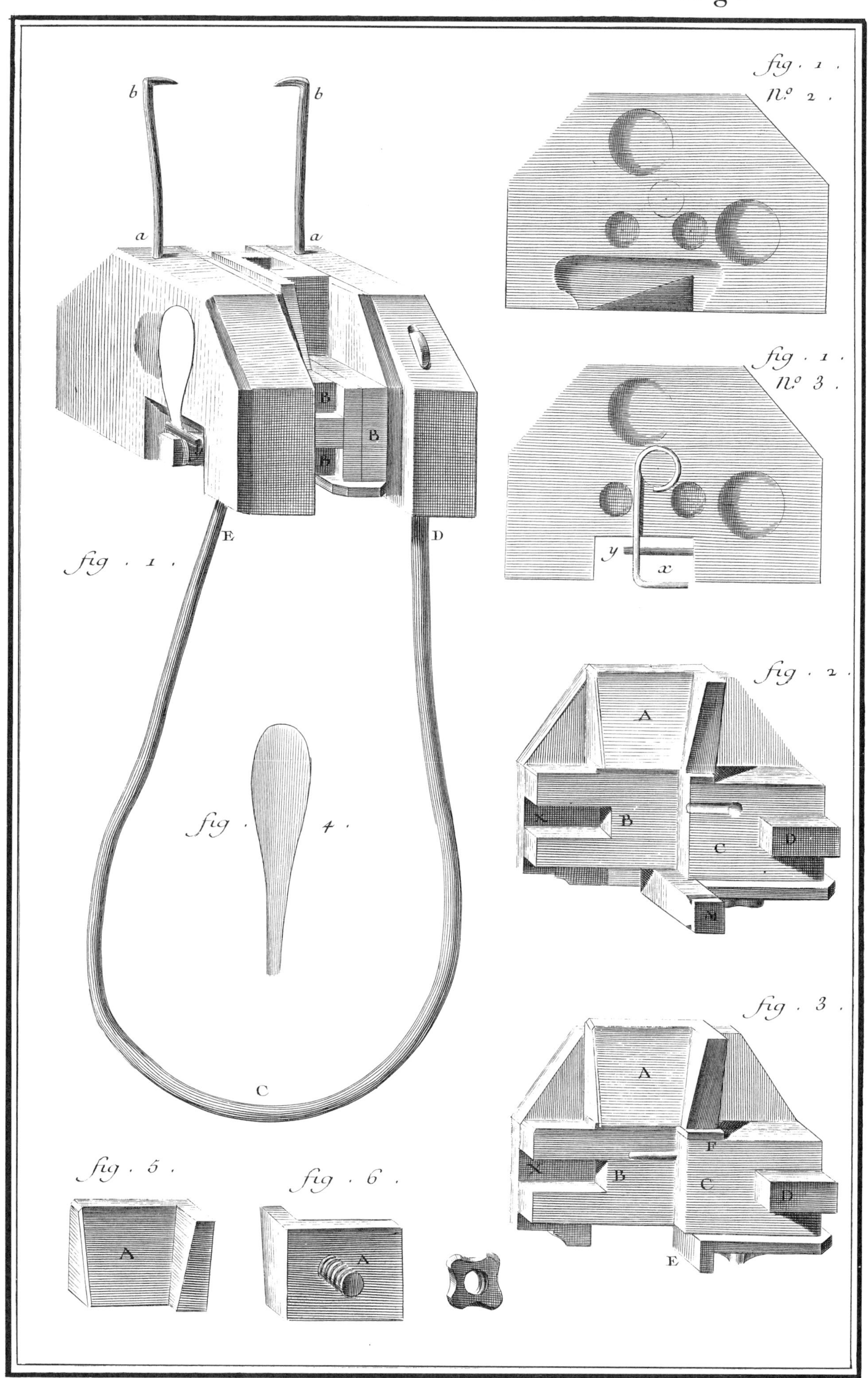
fig. 1.
No. 2.
fig. 1.
No. 3.
y
x
b
b
a
a
B
B
B
E
D
fig. 1.
fig. 4.
C
fig. 2.
A
X
B
C
D
M
fig. 3.
A
F
X
B
C
D
E
fig. 5.
A
fig. 6.
A

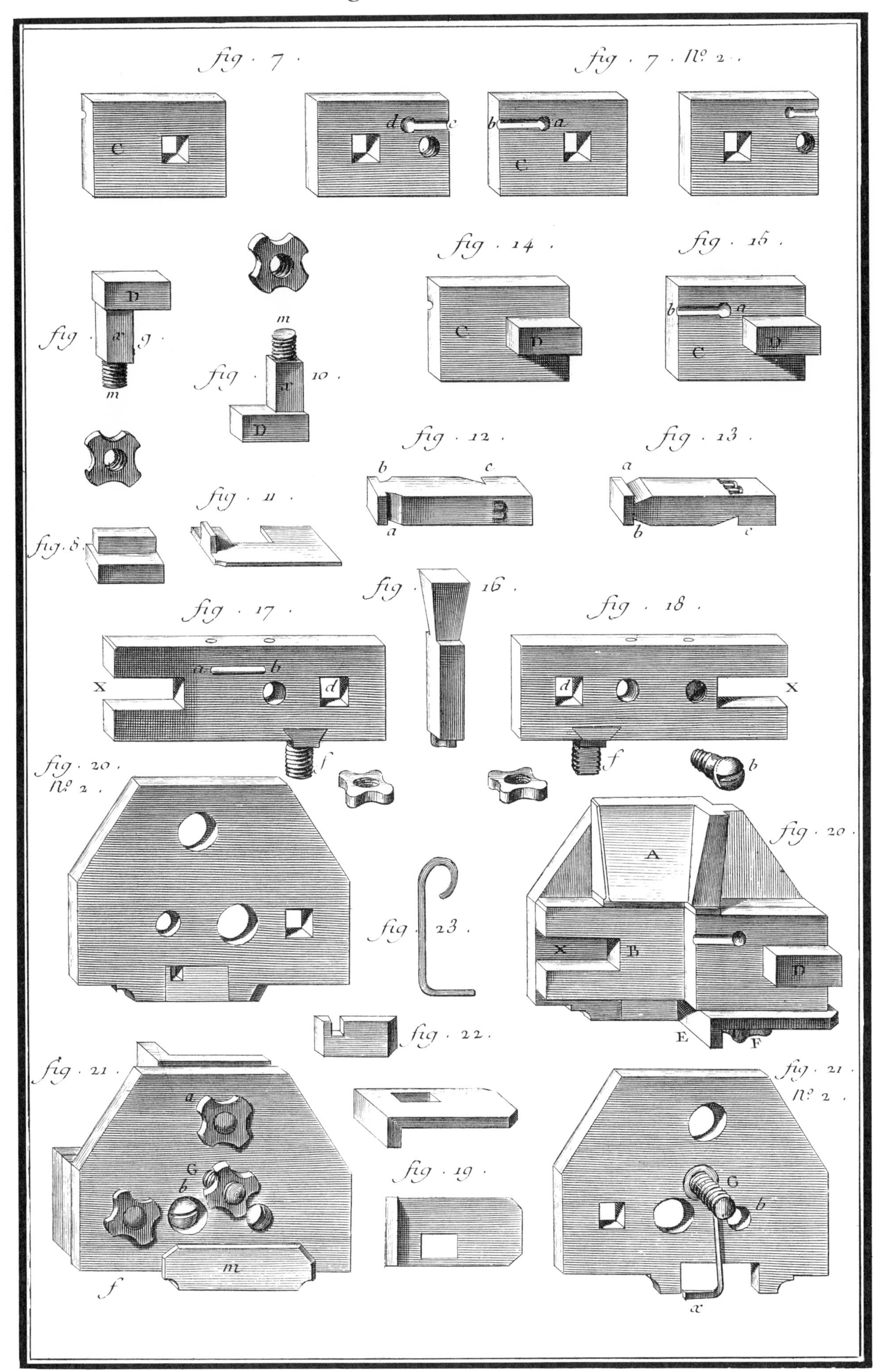
fig. 7.
fig. 7. No. 2.
fig. 14.
fig. 15.
fig. 9.
fig. 10.
fig. 12.
fig. 13.
fig. 11.
fig. 8.
fig. 16.
fig. 17.
fig. 18.
fig. 20. No. 2.
fig. 20.
fig. 23.
fig. 22.
fig. 21.
fig. 21. No. 2.
fig. 19.

Plate 373[1] *Printing V*

In this room of the foundry, the letters are finished. The girl in Fig. 1 "sets up" the type in a "stick," a long narrow frame which clamps them firmly enough so that the boy beside her (Fig. 3) can scrape off the burrs that remain on the foot and the face of each letter. He passes the stick to the finisher (Fig. 2), who fixes it in a vise and planes off the roughness on the bottom remaining at the nub of each tang. The tool scores the entire stick of type, leaving a groove running uniformly up the row of letters.

[1]The editor is indebted to Mr. John S. Carroll for pointing out errors in the identification of this process contained in earlier printings.

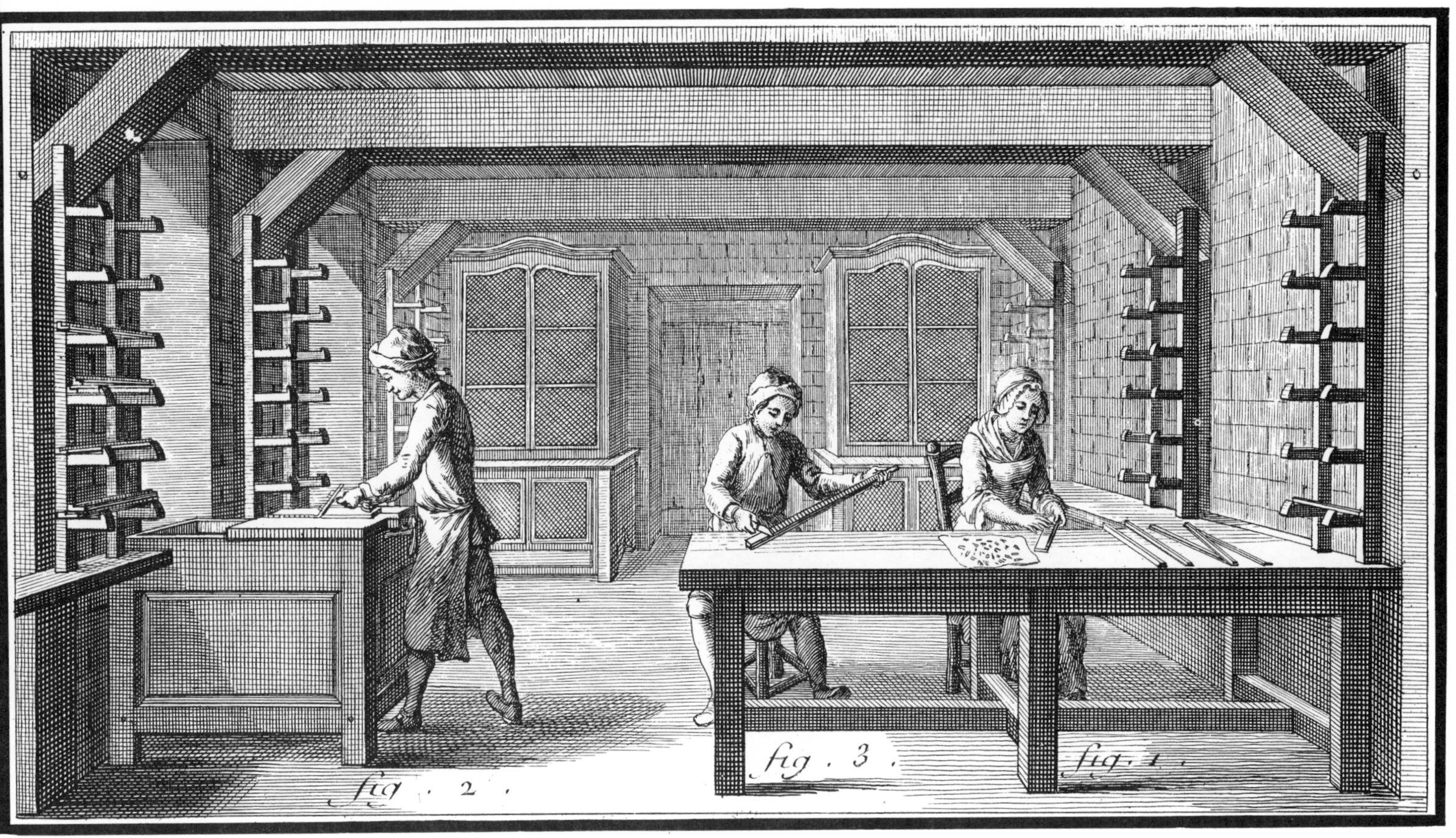

Vol. II, Fonderie en Caracteres, Pl. III.

Plate 374 Printing VI

This is the composing room of a print shop. The typesetter (Fig. 1) has the letters all arranged before him in the typecase. He holds the composing stick in his left hand and picks out letters to set the copy which he has before his eyes fastened to a reading stand mounted on the case. When a line is completed and justified, it is placed in the galley (Fig. 2), a long, narrow tray, from which preliminary "galley proof" is usually printed.

After proof is corrected, the lines of type are made up into pages and the pages "imposed" or arranged so that when the paper is printed and folded, the sequence will be correct. The page of Fig. 3 consists of only two sheets in folio. These have been wedged and locked into the form so that the whole can be lifted and moved about as a solid piece. The worker taps the surface through a block of wood to even the face of the type and assure uniform printing.

The part of the plate above shows typesetters' implements enlarged. Fig. 4 shows the letter "s", a blank space (a), and leads of various thicknesses used to shim out or

Printing VI

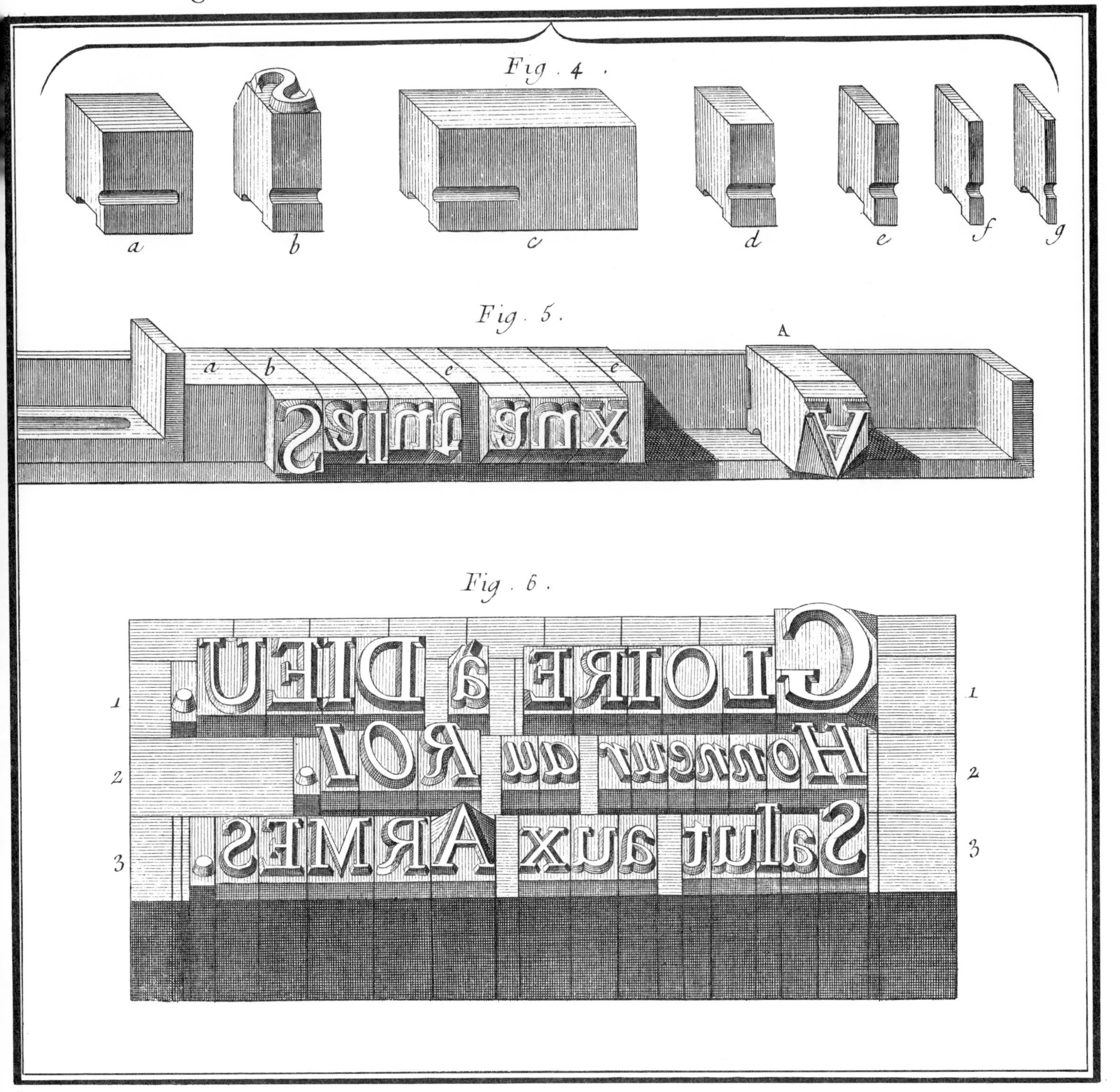

Vol. VII, Imprimerie, Pl. I.

justify the line. Fig. 5 shows a composing stick, with the phrase Salut aux Armes *being set up. An unimpeachable sentiment is being composed:*

Gloire à Dieu.
Honneur au Roi.
Salut aux Armes.

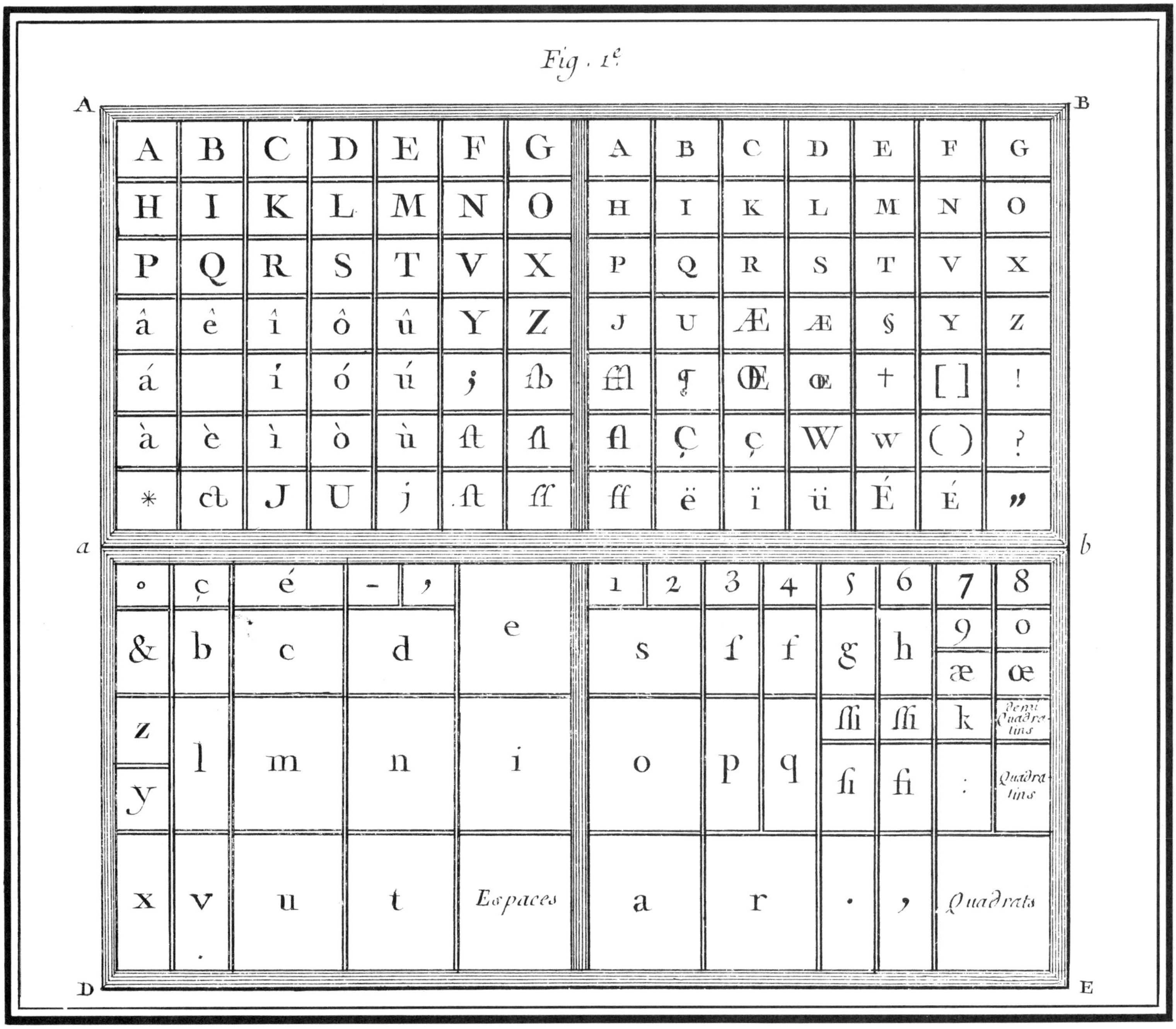

Plate 375 Printing VII

When a frame was printed, the form would be broken up and the type "distributed" back to the case, of which this shows the arrangement. A good man could set up 1500 letters an hour, and distribute perhaps 5,000. This was made possible by the efficient arrangement of the case, which he learned by touch, as a typist learns her keyboard. The upper line is for the capital letters, the lower for small—hence the "upper case" and "lower case" of modern typographical terminology—and the largest compartments are for the most frequently used letters.

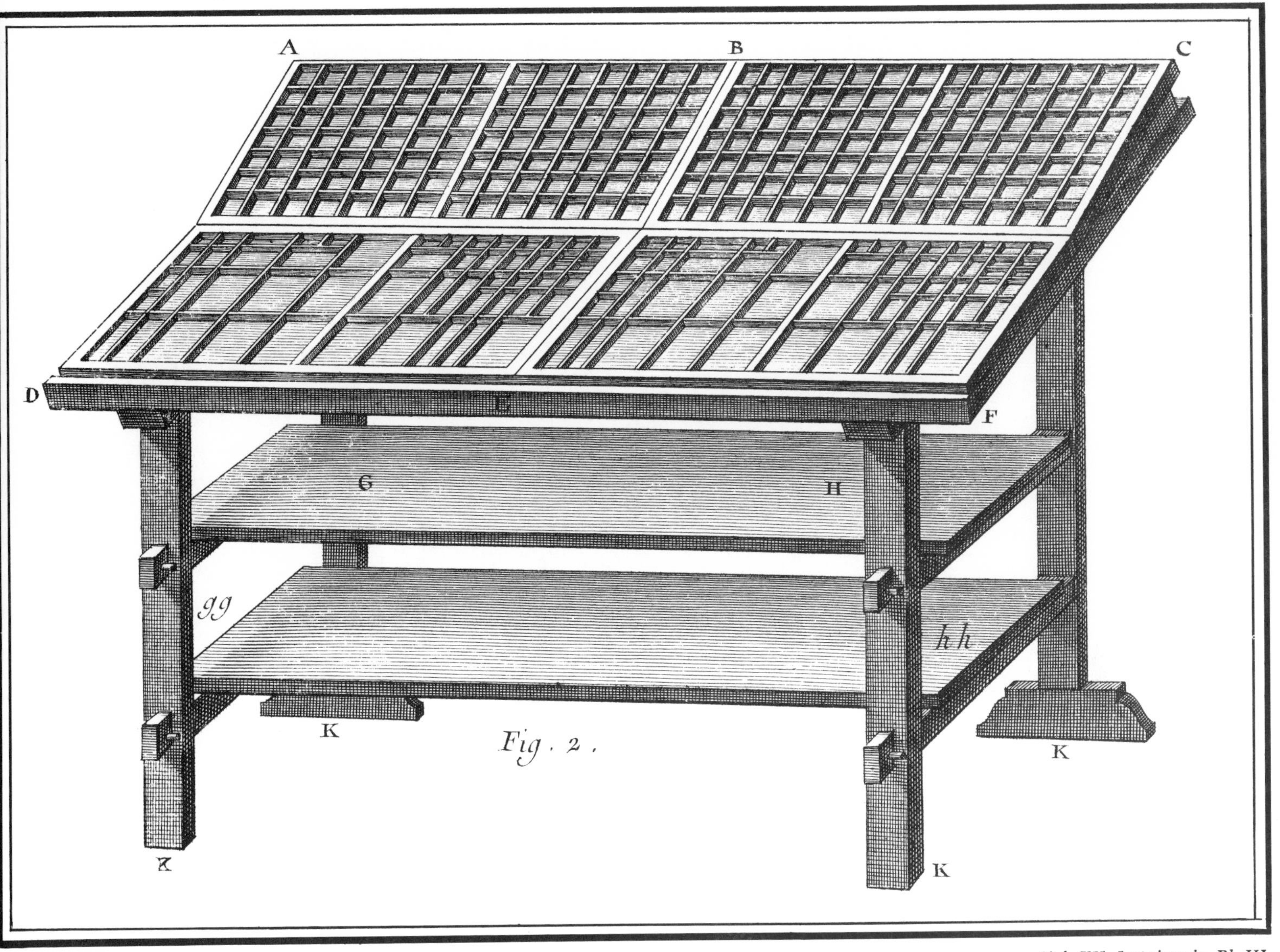

Vol. VII, Imprimerie, Pl. III.

Plate 376 Printing VIII

The same shop serves for washing the forms in mild alkali (Fig. 1) after printing, and moistening the paper before it is printed, to prepare it to take the ink (Fig. 2).

Vol. VII, Imprimerie, Pl. XIII.

Plate 377 Printing IX

This is an establishment with two presses. At the left, one journeyman spreads out a sheet of paper, while his companion (Fig. 2) inks the type. Then when the press is closed, it is slid under the "platen", or upper plate, which is pressed down upon it by means of a screw. A detail of the old hand-press is opposite. The fourth workman has a double job: inking the balls used by (2) and scrutinizing each sheet for even inking and printing as it emerges from the press.

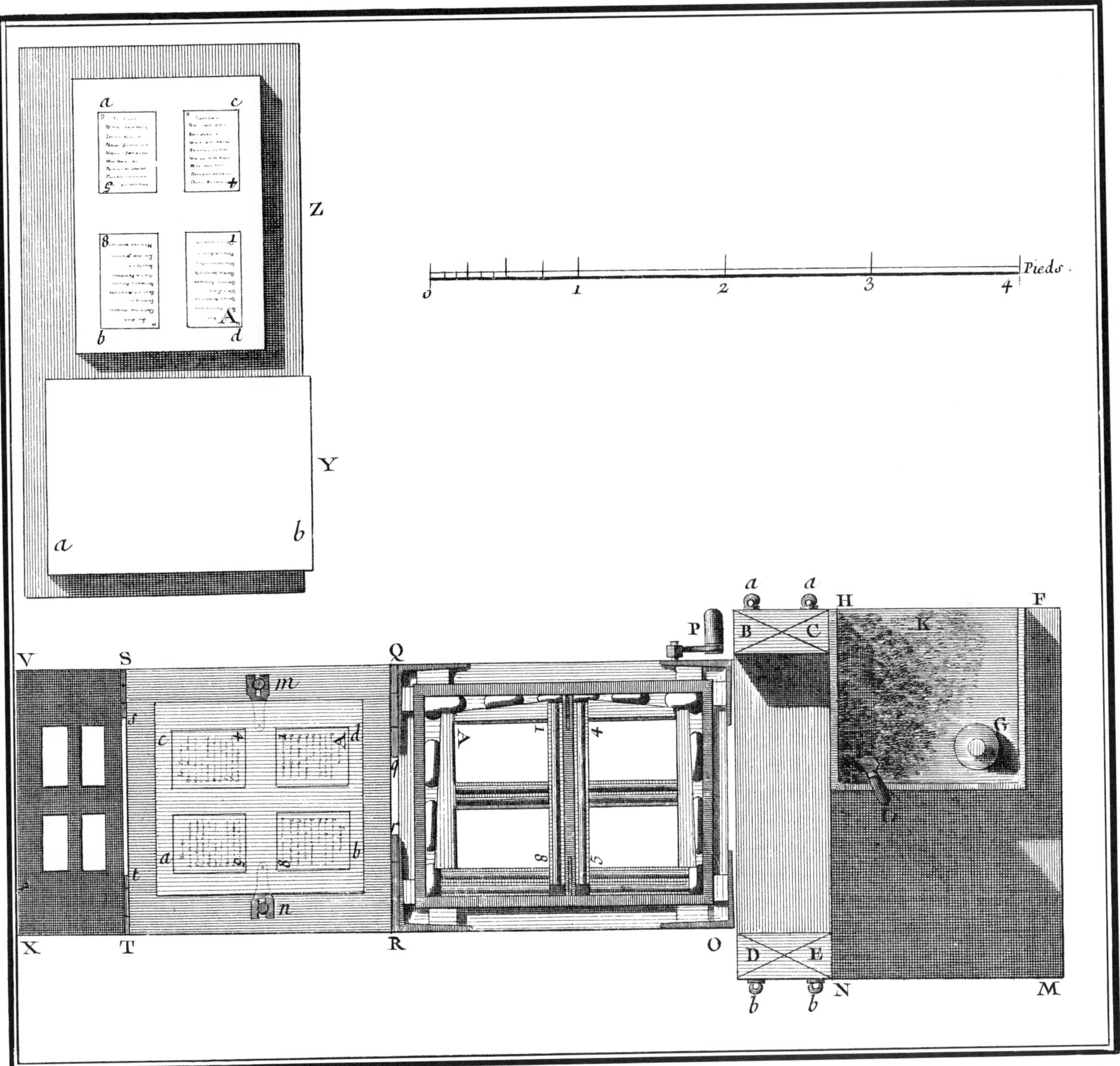

Vol. VII, Imprimerie, Pl. XIV.

Plate 378 Printing X

Now that only bibliophiles know the meaning of "folio, quarto, octavo, duodecimo," etc. up to 64mo, terms which distinguish formats of books according to the manner of printing the pages and folding the sheets, this plate may be of some interest as illustrating these terms. Figs. 1-4 are different arrangements in folio (two pages to a form); Figs. 5-7 are in quarto; and Fig. 8 in octavo. In the lower figures it is clear how the quoins (Fig. 4, m) are used to wedge type into the form.

Fig. 1.
a
b
f
e
g
c
d
Fig. 2.
Fig. 3.
Fig. 4.
Fig. 5.
Fig. 6
Fig. 7.
Fig. 8.

Vol. III, Chaudronnier, Pl. II

Plate 379 Copperplate I

Printing from copperplate was entirely different from letterpress, and formed a separate trade. The plates themselves were produced by the braziers (see Plates 190-191).

Dross and scale are cleaned off by a sharp steel scratching tool (Fig. 1, xx), and the sheet is hammered into plate on a small anvil (Fig. 2). This requires delicacy rather than strength, for the plate must be of uniform thickness. The hammering extends the surface by about a fifth, and it also tempers and hardens the metal. Then the plate is pumiced (Fig. 3) to remove any inequalities, and rubbed with charcoal over a slightly acid bath (Fig. 4). This prepares the surface for burnishing, after which the plate is weighed (Fig. 5), and presented to the engraver who waits with open arms (Fig. 7).

Vol. V. Gravure en taille-douce, Pl. I.

Plate 380 Copperplate II

The engraver's buyer carries his copper plates back to the shop, where the first step is to warm them over a small brazier so that they can be varnished (Fig. 1). In Fig. 1 bis, the varnished surface is smoked, and in Fig. 3 a copier traces the design to be reproduced from the original (l). Figs. 3, 4, & 5 show different methods of handling the acid which etches the design into the metal through the scratches traced in the varnish coating. The more difficult technique of line engraving is illustrated in Fig. 6. Here the engraver cuts the design directly into the copperplate with a burin, a very sharp engraving tool of hardest steel. Either method may have the effect of warping the plate slightly, and if so it must be carefully flattened (Fig. 7) so as to present a perfectly even surface to the paper.

Plate 381 Copperplate III

Although in letterpress the ink is transferred from the surface of the type to the paper, in copperplate, ink is transferred from the grooves and depressions rather than from the surface. First, the plates are inked all over (Fig. a), after which the surface is wiped clean (Fig. b), leaving ink in the grooves. The press (Fig. 1) then forces the paper down into the grooves.

The press, too, is different. It is not a flat but a cylindrical press, which passes the paper and the plate between a pair of rollers.

Typical products of copperplate engraving are hanging on the wall. The original of all the plates in the Encyclopedia *were copperplates.*

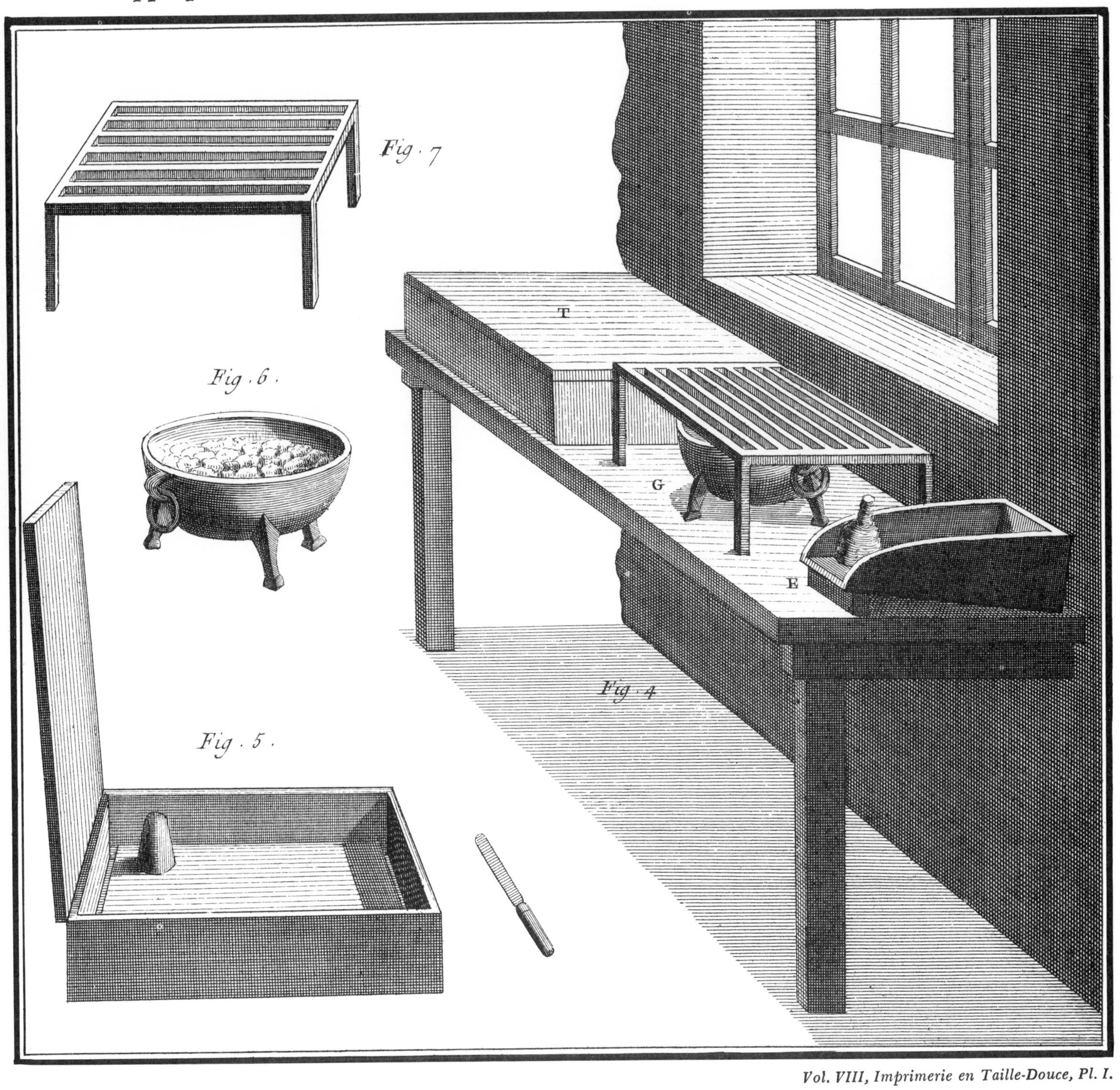

Vol. VIII, Imprimerie en Taille-Douce, Pl. I.

Vol. VIII, Relieur, Pl. I.

Plate 382 Bookbinding I

In France the binding is not yet included in the price of most books. Those who want their libraries between permanent covers engage the services of a binder. Pounding the volume-to-be-bound through a marble block was not a particularly skilled task, and it seems here to have been confided to one who is the epitome of an "occasional" laborer (a). Stitching signatures was a woman's work, for which she used a special frame (b). After binding, the pages are cut (c). A number of volumes are put under compression for a time in the big press to prevent warping (d).

Plate 383 Bookbinding II

A really sumptuous binding would be set off by gilding, applied to the edge of the pages (a) and to spine (c) and cover (b). The artisan working on the cover presses a design into the calfskin, which he will fill with gold leaf. Gold leaf comes from the gilder (see Plate 418) in little booklets (Fig. 2 C, Fig. 10). The press of Fig. 5 holds the volumes while the edges are gilded.

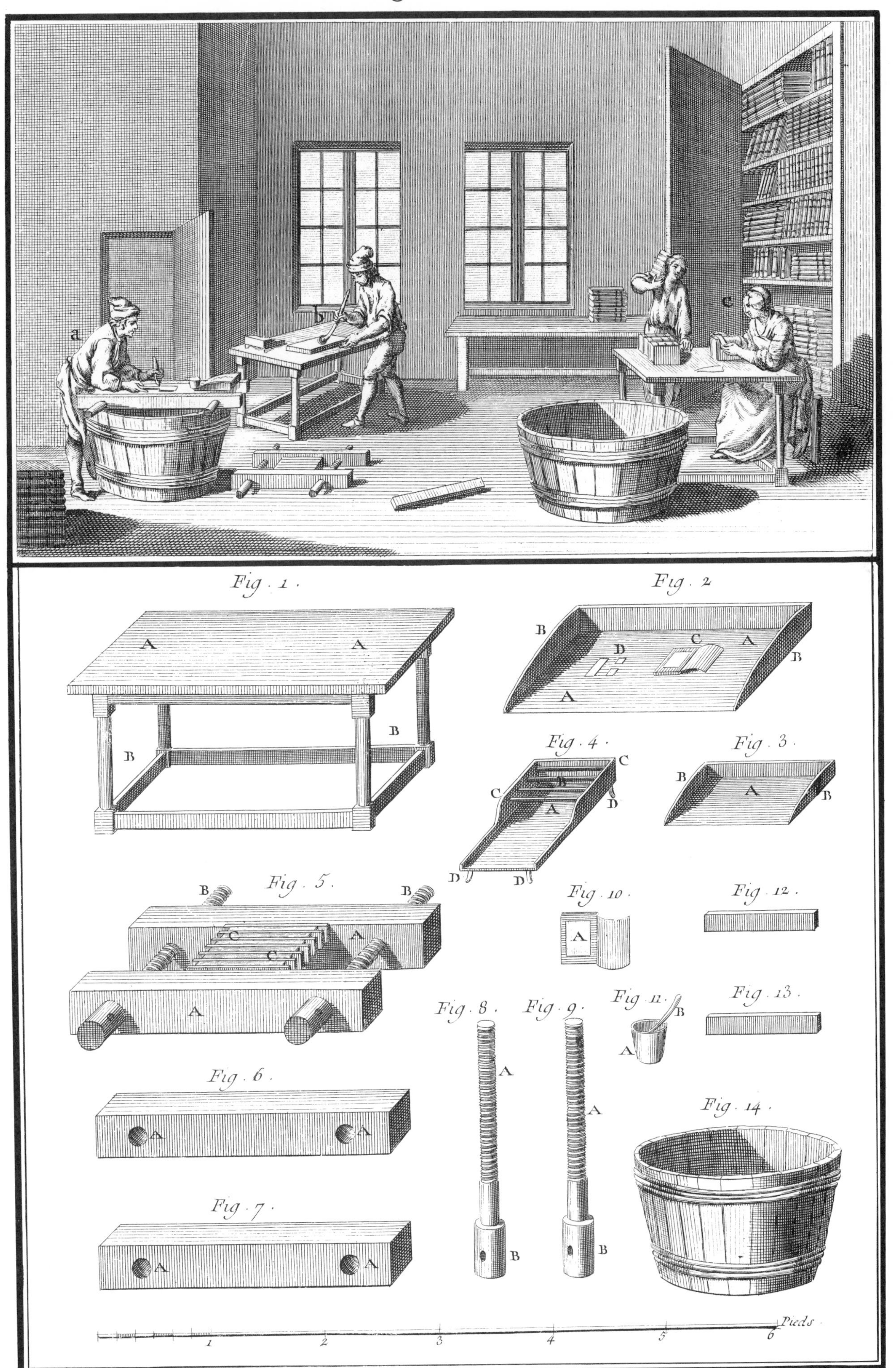
a
b
c
Fig. 1.
Fig. 2
Fig. 3.
Fig. 4.
Fig. 5.
Fig. 6.
Fig. 7.
Fig. 8.
Fig. 9.
Fig. 10.
Fig. 11.
Fig. 12.
Fig. 13.
Fig. 14.
Pieds

Vol. V, Marbreur de Papier, Pl. I.

Plate 384 Marbled Paper I

Marbling paper consists in coloring it to get by art the mottled effects that result from accident when a piece of paper is dropped into the water. "The Lebretons, father and son, who worked at the end of the last [seventeenth] and into the present century, achieved real masterpieces in this genre. They had the secret of mingling threads of gold and silver with colored waves and veins in paper, and it was truly remarkable, the taste, variety, and richness which they introduced into work essentially so frivolous." And the Encyclopedists, who prided themselves on getting the best technical information from the artisans themselves—no easy matter particularly when it concerned one of these tiny trades which thrived on secrets—went to the widow of Lebreton fils, *who had fallen into penury and distress, to learn about the art.*

The idea of the process is to float colors on gummed water, as cream can be floated on coffee, and then to touch sheets of paper to the swirled rainbows of the surface. The worker of Fig. 1 is putting through a sieve piny water in which gum has been soaking (e) for three days. Across the shop another pulverizes colors—indigo, brazilwood, ochre, lamp-black, which are cast or sprinkled on (Fig. 3) the gummed water in a shallow pan. First there goes blue; then red, which pushes back the blue here and there and takes over spots and areas for itself; then yellow, green, black, each color

Plate 385 Marbled Paper II

Vol. V, Marbreur de Papier, Pl. II.

staying to itself. Now art imposes itself on chance and the mottled surface of the liquid is gently combed (Fig. 4) to achieve the graining and massing of colors desired by the artist. This, finally, is transferred to paper by touching the sheet to be marbled (Fig. 5) to the surface for a moment or two till just wet through, after which the papers are drained (Fig. 7) and dried (Fig. 8).

Plate 385 Marbled Paper II

Marbling was used chiefly to dress up books in handsome boards or endpapers. The dried sheets were finished by waxing (Fig. 9), polishing (Fig. 10, No. 1), and folding (Fig. 10, No. 2). The surface presented by the edges of the pages in a volume might also be marbled. That is the occupation of the artisan of Fig. 11, No. 2, who spreads the covers and holds the edge of the book to the painted water.

Plate 386 Cardboard

In the 18th century cardboard was not the rough-finished, shoddy material which goes under that name today. Its chief uses were in the manufacture of playing cards and book-boards, the latter to be covered by the binder in marbled paper, morocco, or calf. The technique of fabrication was adapted from papermaking. The chief differences were in the mold, which dipped a thicker sheet of pulp from the bath, and in the raw materials. Cardboard was made from scrap paper rather than directly from rags.

The process began in an adjoining shed where piles of old paper and used cardboard collected by the junkman were set to fermenting in moist piles. What with the damp heat of the shop and the overpowering smell of mildew, the atmosphere of that shop was almost insupportable. After a week or so, the soggy mass was shredded by hand in great troughs and pulped to a homogeneous suspension in the circular horse-powered mill of the vignette. Each lot would be macerated for an hour or two, "depending on how fast the horse moves."

Thereafter the pulp is transferred in buckets to the dipping trough, from which the workman is just lifting a mold. He lets the excess moisture drip into the drainboard (CD) arranged on sawhorses while he takes the previous sheet (G) and turns it out of the chain-mold onto a pile to be pressed (HI). The bottom sheet fits into a pressing frame (Fig.10 opposite), and successive sheets are separated by oaken strips of the same size. When the pile reaches a depth of 120 sheets, it is moved bodily to the press (AB) and squeezed almost dry. The cardboards have then to be cleaned, trimmed, and tho-

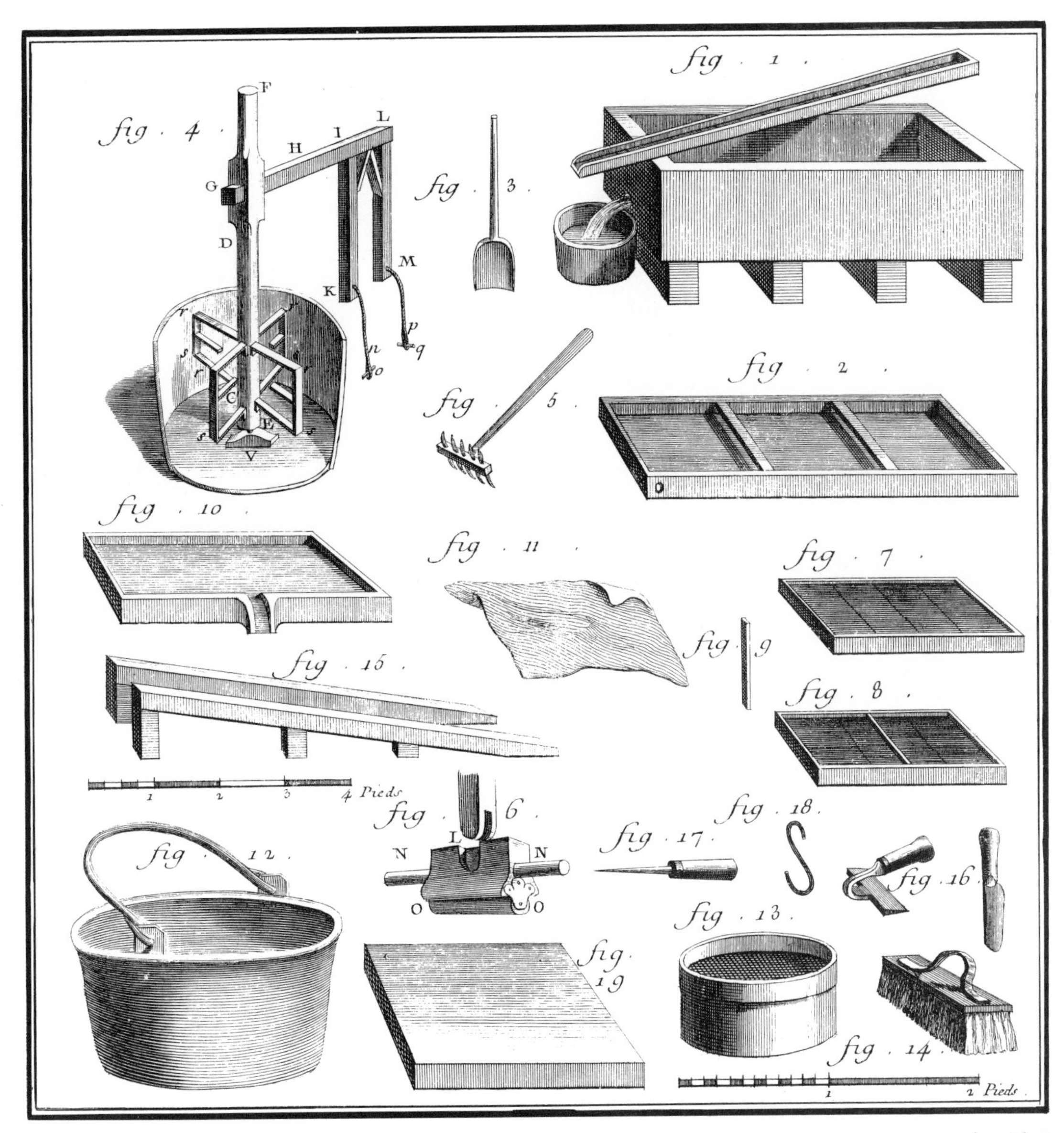

Vol. II, Cartonnier, Pl. I.

roughly dried in an airy attic. The thickness of the final product may be adjusted by omitting the dividing strips from the pile, in which case the press laminates successive sheets after the manner of modern plywood, except that there is, of course, no cross-graining.

It may, perhaps, be worthwhile to reproduce the tools of the trade. This plate is a good illustration of the detail and clarity of which the Encyclopedia *was capable as a technical dictionary. It would be possible to read the account, look at the illustrations, and go into the business. The Encyclopedist may have had some such hope in mind. "France does a considerable trade in cardboards. But I have visited many workshops which I have not found as well run as the one just described. It seemed to me that the artisans did not bring to their work as much attention or neatness as they might have done. It is by no means the only occasion I have had to notice that, provided the work gets done, no one worries much about how well."*

Leather

Vol. II, Boucher, Pl.

Plate 387 The Butcher

A section on leather may appropriately begin with the work of the man who, in a sense, produces hides, the butcher.

Vol. III, Boyaudier.

Plate 388 Manufacturing Gut Strings

One of the least agreeable of Parisian trades was that of gut-spinner, who supplied the tennis racket manufacturer and violin-maker with string. The entire guild consisted of eight masters, all of whom had establishments near the slaughterhouses of the Faubourg St. Martin. They would obtain sheep gut and lamb gut from the slaughterer, press out the ordure, wash and dry the lengths of gut on racks (Fig. 1), then scrape it (Fig. 3), and splice it (Fig. 4). It could be sold as a single strand for the thinner strings on a violin, or spun into cords of several strands for a bass viol. In English this material is called catgut, but it has in fact always been prepared from sheep and goats.

Plate 389 Tanning I

Like the paper factory of l'Anglée, the Encyclopedia's *tannery may be taken as an example of the degree of rationalization of which 18th century industry was capable. It is a particularly significant example, perhaps, since tanning hides for leather was an old and traditionally a messy process.*

A preliminary glance at the plan will make the sequence of operations clearer. Of necessity the establishment was built along the river bank in a connected series of warehouses and sheds occupying an area of about 75 by 200 feet. There were four main divisions to the plant: a washing platform along the river (GHIK), a liming shop with four cisterns (QRST), a larger shop containing fourteen tubs for rinsing and soaking, and a courtyard containing 12 tanning cisterns. A stairway (Y) leads to the attic used for drying leathers. Facing on the street were rooms for oiling and finishing leathers (h, f, d). Across the entry-way was a lodging (m, l, k) for the master of the tannery, who lived right on the job.

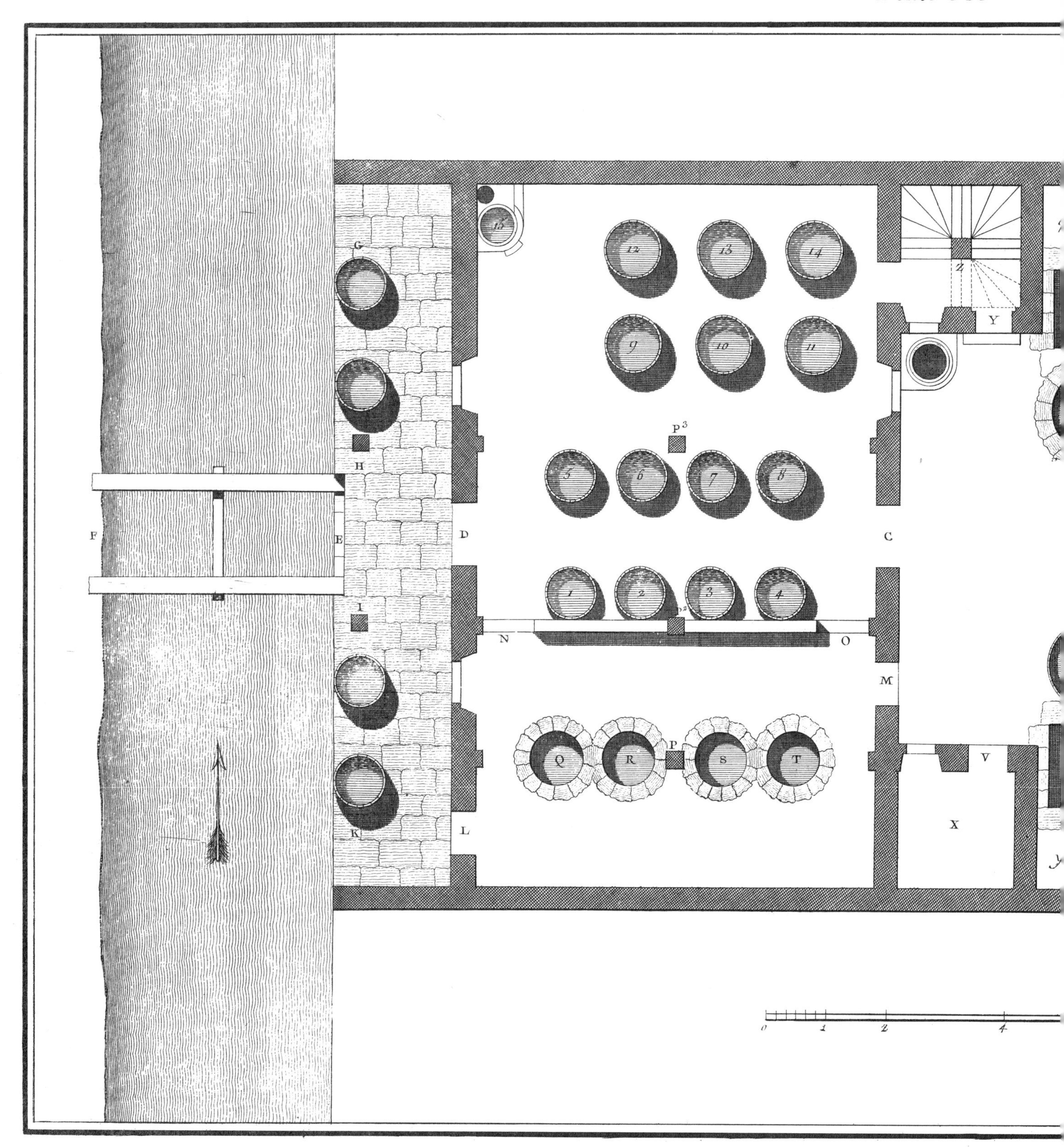

G
H
I
K
F
E
D
L
C
M
N
O
P
P³
Q
R
S
T
V
X
Y
Z
1
2
3
4
5
6
7
8
9
10
11
12
13
14
15
0
1
2
4

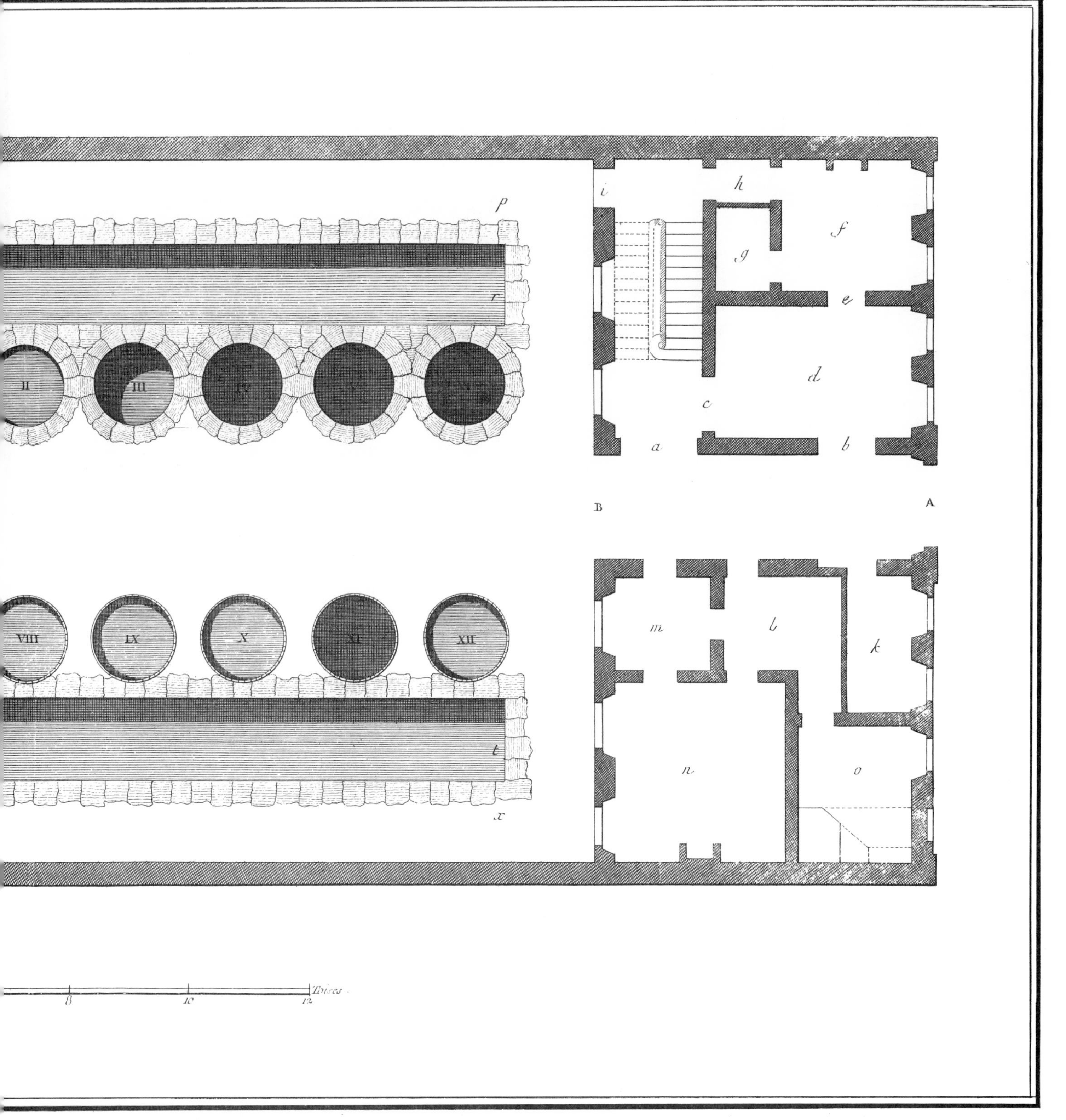
p
r
II
III
IV
V
VI
VIII
IX
X
XI
XII
t
x
i
h
f
g
e
d
c
a
b
B
A
m
l
k
n
o
8
10
12
Toises

Vol. IX, Tanneur, Pl. III.

Plate 390 Tanning II

The work along the river bank consisted, first (Fig. 1) of washing off the grime, blood, and loose bits of flesh adhering to hides. The slaughter house had simply stripped the skins rudely from the carcass. After cleaning, the hides were put to soak until ready for liming inside the tannery.

Liming loosened the hair and began the curing process. (See Plate 391). After being steeped in lime-cisterns, the hides were then brought out again to the water's edge to be scraped (Fig. 3). The worker used a blunt tool to press out the lime water. The paving of this work area is slanted to permit rinsing the refuse into the stream. The hair, however, was collected to be sold to upholsterers.

In any city where tanning was practised, there is still likely to be a picturesque and smelly Rue des tanneurs *along the river. Tanning was one reason (dyeing was another) that only in modern districts have riverside properties been thought desirable. In pre-industrial times, the river area was more often what the railroad district is in American towns.*

Vol. IX, Tanneur, Pl. IV.

Plate 391 Tanning III

Liming hides began in a solution of old lime (Fig. 1) from previous runs. Though weak, this had the advantage of containing much rotting organic matter, and the bacteria attacked the hair roots in a gentle but determined fashion.

Hides were left to work in old lime for two or three days, then folded in a wet pile for another four or five, then returned to the cistern for another soak, and so on for about two months. After that, the hair could be scraped off easily without tearing or marring the skin.

Once unhaired, hides had to be limed again, this time in a weak solution of new lime, in which they were alternately soaked for three or four days and stacked for a week for a period of three months. Nor was the liming finished even then. The final stage consisted of alternate weekly soakings and stackings in strong lime for four months.

Vol. IX, Tanneur, Pl. V.

Plate 392 Tanning IV

Hides might also be prepared for tanning by curing in a leaven of barley water. The method was far less tedious than liming, but was suited only to hides of the better grades. The hides had to undergo soaking in four progressively stronger concentrations. They needed only two or three days in each, but they had to be turned and drained twice every three hours.

The other part of this warehouse, which lies to the right and is not shown, serves for preliminary soaking in tannic solutions (9, 10, etc). After this the hides were ready for tanning proper in the yard.

Plate 393 Tanning V

Vol. IX, Tanneur, Pl. VI.

Plate 393 Tanning V

The courtyard contained a well in one corner. Parallel ditches and tanning pits ran down each side. Two workmen bring out a pile of hides to be laid away (Figs. 1, 2) and a third a pack basket of tanbark (Fig. 3).

The bottom of a pit is covered first with a layer of used tanbark, on which is spread an inch of new bark. On that is laid a hide, then another layer of tanbark, then a second hide, and so on (Fig. 4) until the pit is full (Fig. 5) to within two feet of ground level. Finally a foot of old bark is spread across the top, the whole is trampled down to compress it as closely as possible, and the hides are left in pits with continual moistening for three months.

What with curing in lime and tanning, therefore, it took almost a year to process a single hide. When the time came to unpack the pits, the hides would be all soft and pliant and the workers had to handle them carefully. The man in the pit (Fig. 5) worked barefoot, and tossed the used tannin back into the ditch behind him (Fig. 7), where another workman made pellets of it (Fig. 8) to be stored for re-use.

Vol. III, Corroyeur, Pl. I

Plate 394 Tanning VI

After leaving the tanner, leather was still not finished, but had to be curried. The currier impregnated the leather with oil or grease, and readied it for the use of the bootmaker or harnessmaker or other craftsman.

The hides were first oiled and trodden (Fig. 1), then scraped with a variety of knives and combs and buffers (Figs. 2 to 5), and finally pommeled with a curious four-headed hammer (Fig. 6).

Vol. III, Hongroyeur, Pl. III.

Plate 395 "Hungarian" Tanning

Curing hides with tallow was known as Hungarian tanning. This is the final step. On the rack (G) are cowhides, which have undergone washing, scraping, soaking in brine and alum, and pounding. Molten tallow is then brushed into them (Figs. 1, 2) and the hides are singed over an open brazier (Figs. 3, 4), so as to seal the pores in tallow while melting off the surplus. These "Hungarian" tanners had to work in an airtight room where their own pores responded like those of the hides.

Vol. II, Chamoiseur et Megissier, Pl. I.

Plate 396 Chamois I

Sheep and goats yielded a lighter leather than did cows or oxen. The best grade, oddly enough, was produced from strains of sheep with poor wool. Baby lambs were killed for kidskin, while foetal lambs produced the best leather for gloves. Goatskin was tougher. Its tight texture and hard grain suited it to fancy work of all descriptions. Both sheepskin and goatskin were processed into chamois, used for fine garments, rather than for cleaning cloths. A pair of chamois breeches was the height of comfort.

This is an establishment for curing sheep and goat skins. The master procured his hides from the slaughterhouse, all matted with blood and filth. A superficial washing (Fig. 1) was followed by thorough washing, scraping, and draining (Fig. 1, No. 2; Figs. 2 and 3), which was repeated three times.

Curing was by lime. The workman made a pile of skins stretched out hair side down. One after another the fleshy surfaces were coated (Fig. 5) with a solution of lime from a bucket, the workman using a swab of sheepskin. As each surface was covered, it was removed, folded upon itself lengthwise, and laid in a second pile (Fig. 6), where hides stayed for two days. But if the skins had first been dried, they had to be limed for a week to ten days. Then after the lime had done its work, the softened and cured hides were rinsed in the river, sunned for an hour, and trimmed of any hair spoiled in the curing (Fig. 4).

Plate 397 Chamois II

Vol. II, Chamoiseur et Megissier, Pl. II.

Plate 397 Chamois II

After the hide had been limed and sunned, the wool or hair became so loose that it could be removed by scraping (Fig. 1).

Other scrapings, washings, wringings and beatings then followed (Figs. 3, 4, 5, 6, 2) to remove every trace of flesh that still clung to the skins. Even this did not finish the process. A further laborious liming and soaking was still necessary, and was carried out in the tubs shown (Figs. 7 and 8) in the preceding plate. Altogether it took about three months to process a sheepskin or goathide into the soft and pliant chamois of commerce.

Plate 398 Morocco I

"Morocco" leather no longer had anything to do with that kingdom in the 18th century. The process had spread throughout the entire Mediterranean world from its origin in Moorish Spain and North Africa.

Genuine morocco was made from goatskins. The distinctive feature of the process was that the skins were dyed before tanning. These are the preliminary preparations: scraping (Fig. 1) and soaking in great cisterns about five feet deep, (Fig. 2, A, B), in which skins are handled by great pincers. After thorough soaking, the hides are further softened by kneading them (Fig. 4) under water with pestle-like iron rods.

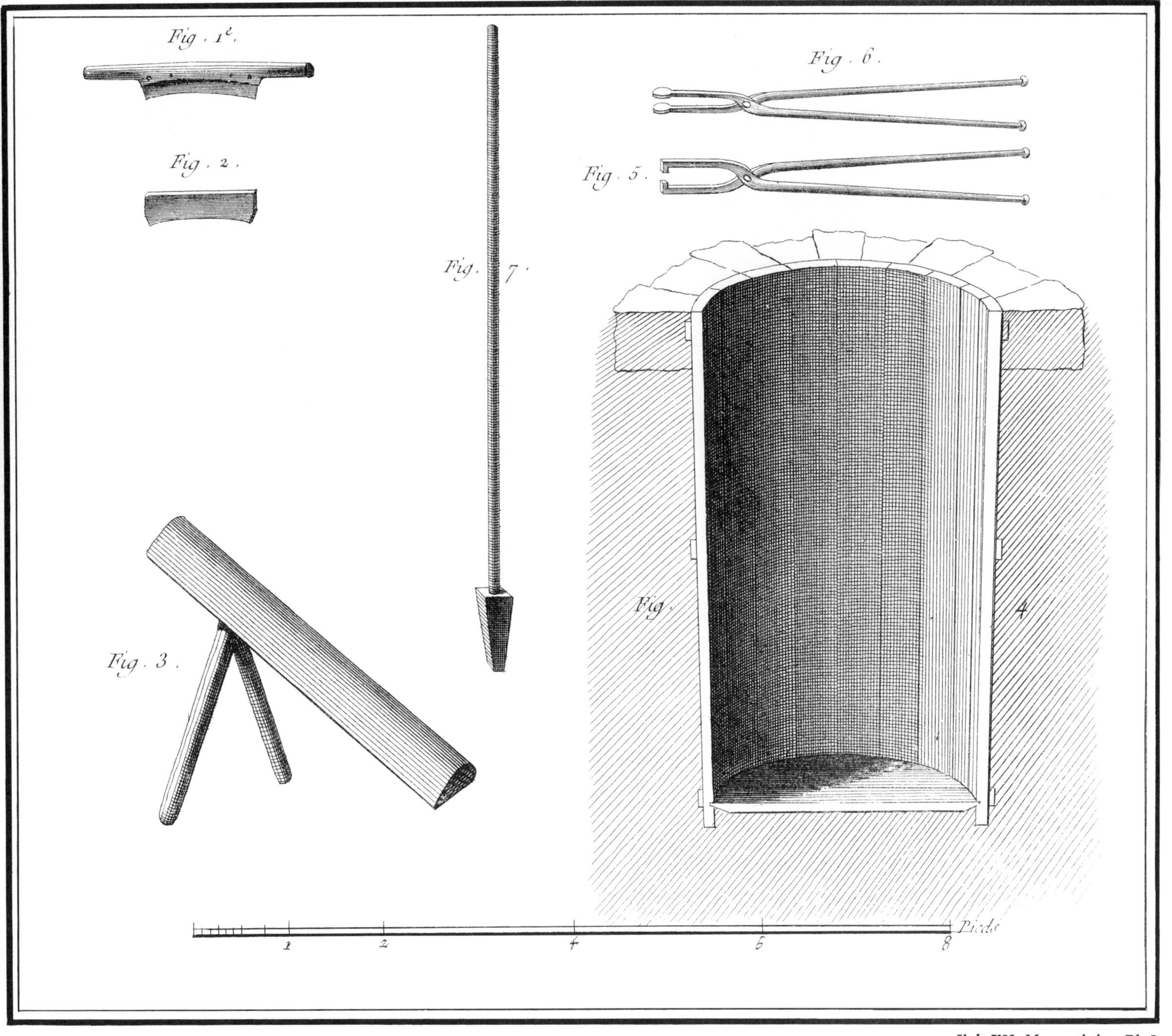
Fig. 1e.
Fig. 2.
Fig. 7.
Fig. 6.
Fig. 5.
Fig. 3.
Fig. 4
1
2
4
6
8
Pieds

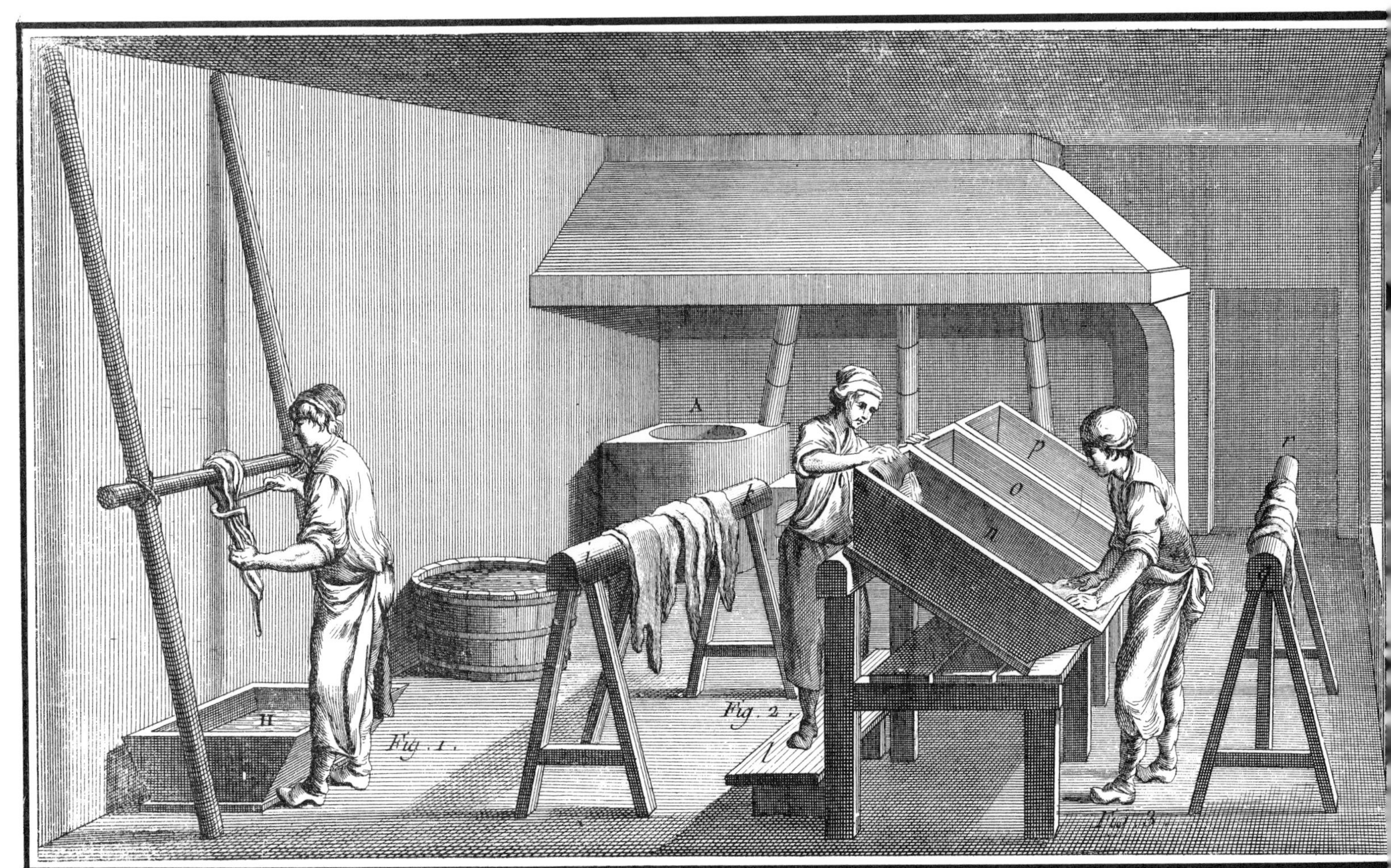

Vol. VII, Maroquinier, Pl.

Plate 399 Morocco II

In the dyeing shop, hides are wrung out (Fig. 1) and folded flesh side in across a clothes-horse (Fig. 2). Moroccoed leather was dyed on one side only, the outer side. The dye was "kermes," (made like cochineal from small dried insects). With alum as a mordant, it gives the brilliant red characteristic of morocco. It was prepared in the cauldron at the rear. In the foreground one worker adds dye to the bath (Fig. 2), through which the dyer passes the folded skin two or three times (Fig. 3).

After tanning, skins treated in this way took on the soft yet fine-grained texture and the clarity of color for which they were prized in the binding of fine editions, or in ladies' belts, or in the many other uses of ornamental leather.

Vol. VIII, Parcheminier, Pl. I.

Plate 400 Parchment

Parchment still found considerable use in the 18th century, for binding books or for ceremonial use in charters, diplomas, and other documents. Most manufacturers bought their sheepskins already cured—that is washed and limed—from the tanner, after which their work differed from the tanner's only in the necessity of producing a more even, finely grained surface. But the operations are not essentially different: unhairing and scraping (e); soaking and washing (c, a), stretching as taut as possible on a frame (d); chalking and pumicing out the smallest inequalities (b); and lastly cutting to the prescribed size (f). The old trades embodied a great deal more of this sort of monotonous work requiring great care but little skill than is always appreciated.

Gold, Silver & Jewelry

Plate 401 Silver & Goldsmithing I

This might be either a silversmith or a goldsmith, though it is more likely that silver was the metal being worked. In either case, the master would be a member of the corps or guild of jewelers, one of the most highly privileged of the companies of Paris, specializing in Orfèvre grossier, *that is to say, in plate.*

Jewelers formed one of the six senior guilds of the City of Paris. Their statutes dated from 1260 and limited the number of masters to three hundred, the most fortunate of whom were quartered in the galleries of the Louvre. No one might be admitted to the trade without having served an apprenticeship of eight years, after which the candidate had to submit an original chef d'oeuvre *to the critical examination of a jury—except that there might always be four masters licensed by the King and two by the Duc d'Orléans, and every eight years the orphanage of the Trinity had the right to name two masters whom it had apprenticed from its roster. Widows might carry on their husbands' shops, but not take apprentices.*

Such were a very few of the provisions which guarded and hedged about the trades in the old corporate regime before the Revolution. The Encyclopedists are often described as relentlessly opposed to this order of things and dedicated to the competitive tenets of economic liberalism. So in general they were, but not in as sectarian a fashion as is

Vol. VIII, Orfèvre grossier, Pl. I.

often said. For they were Frenchmen and believed in the imposition of standards. The author of the article on the goldsmiths took issue with the liberal views on economics which he thought irrelevant to art. Only by the most careful training, he felt, only by the most rigorous regulation and supervision, would it be possible to maintain that dominion of taste, skill, and craftsmanship which made Paris the jewelry capital of Europe.

Plate 402 Silver & Goldsmithing II, III, IV

As will be apparent from the next few plates French taste would never have been satisfied by the chaste productions of a Paul Revere, nor even with the work of English silversmiths of the 18th century. As in other fields of decoration, French styling leaned more toward the baroque.

Vol. VIII, Orfèvre grossier, Pl. IV.

Vol. VIII, Orfèvre grossier, Pl. V.

Vol. VIII, Orfèvre grossier, Pl. VI.

Plate 405 Silver & Goldsmithing V

Readers interested in machinery may wish to see something of the precision tools of which the 18th century was capable. This is a goldsmith's lathe, on which he is turning a platter for a gold service.

Vol. VIII, Orfèvre grossier, Pl. XVII.

Plate 406 Silver & Goldsmithing VI

An enlargement of the lathe. Fig. 11 is the platter, which is turned by the rim of Fig. 10, the holes of which match up with those on the main turntable (Fig. 9).

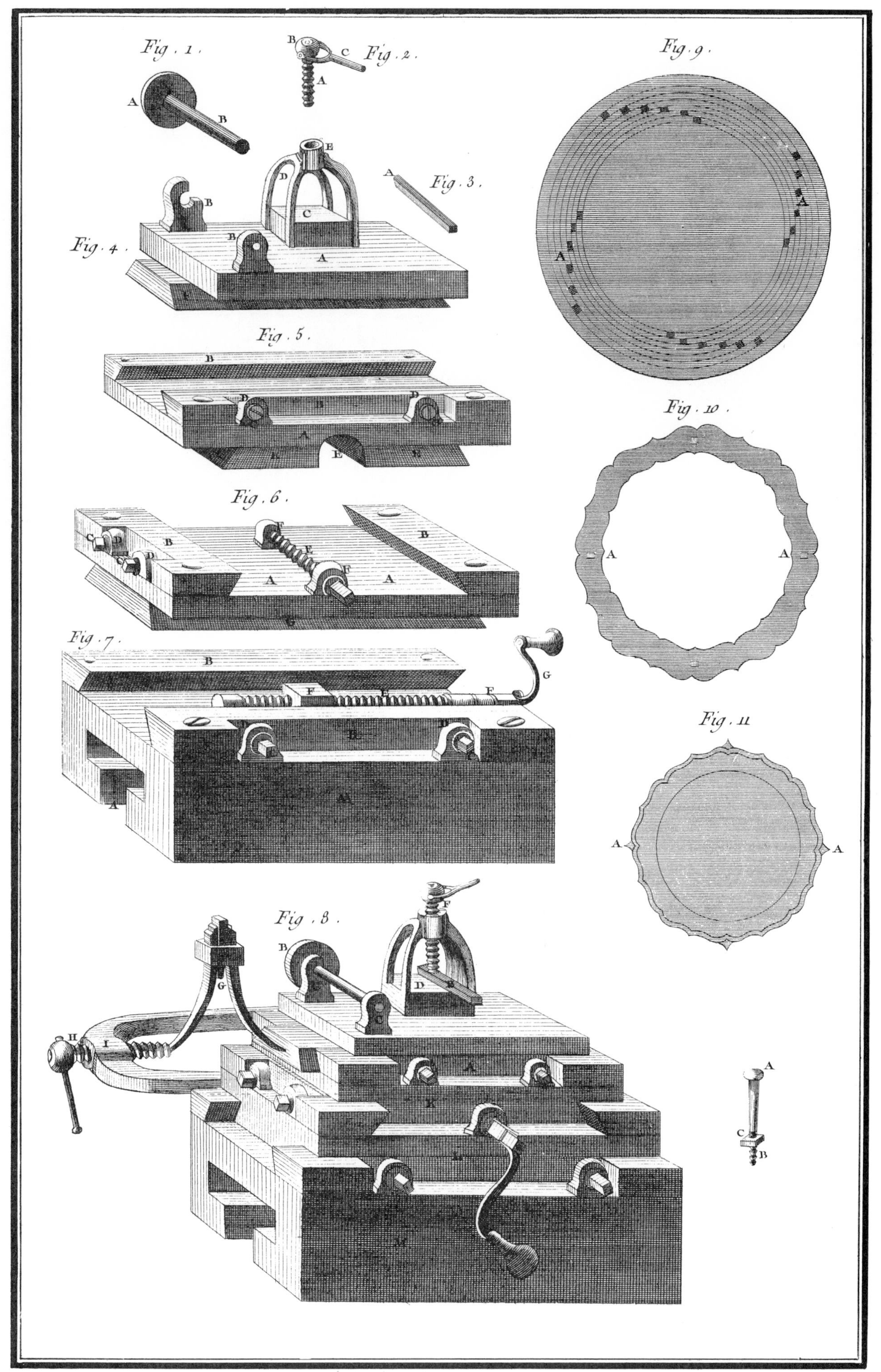
Fig. 1.
Fig. 2.
Fig. 3.
Fig. 4.
Fig. 5.
Fig. 6.
Fig. 7.
Fig. 8.
Fig. 9.
Fig. 10.
Fig. 11.

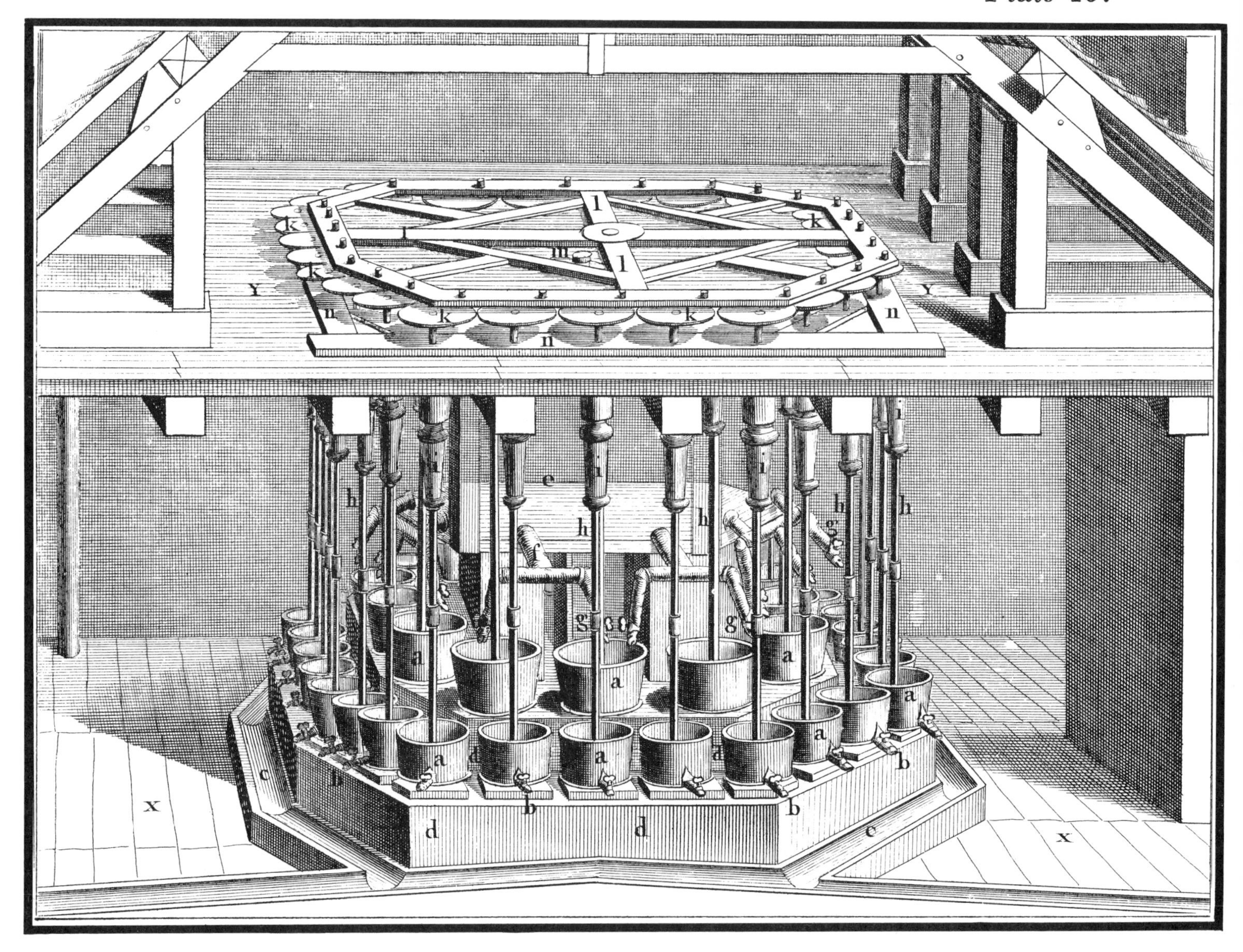

Plate 407 Silver & Goldsmithing VII

The machine above serves for washing gold; that opposite is a mechanical mortar and pestle built to the scale of the goldsmith's requirements.

Vol. VIII, Orfèvre grossier, Pl. XIX.

Vol. VIII, Orfèvre Bijoutier, Pl. I.

Plates 408, 409, 410, 411 The Jeweler I, II, III, IV

Closely related to the goldsmith's art was the jeweler's. His forge is even smaller. The characteristic work-counter with its pockets underneath each place may still be seen in the shops of Parisian jewelers. The sales counter is decorated in a manner befitting the trade, and Madame herself handles the scales and quotes a price which she is, no doubt, prepared to be reasonable about. The following plates illustrate some of her wares.

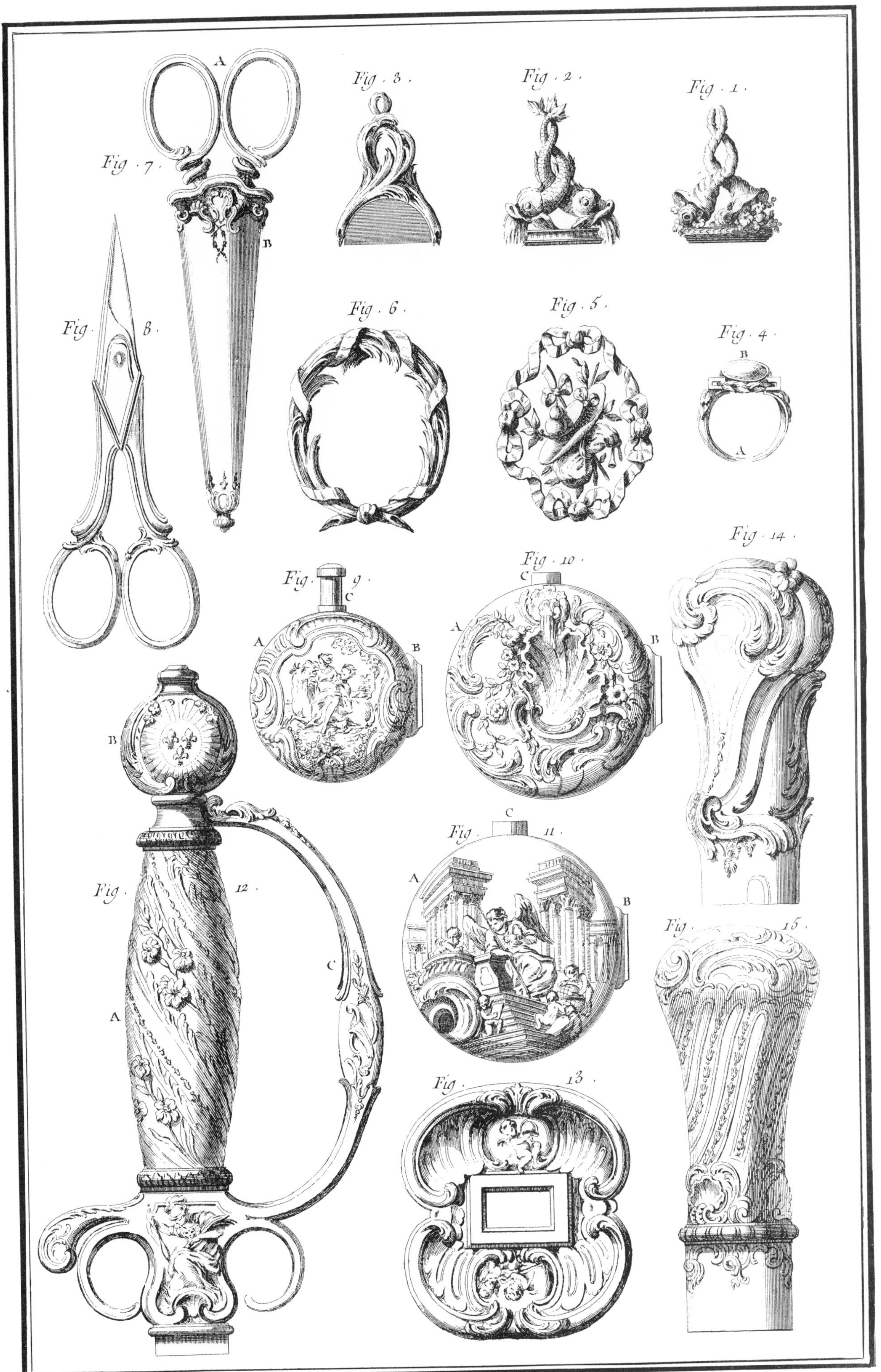
Fig. 1.
Fig. 2.
Fig. 3.
Fig. 4.
Fig. 5.
Fig. 6.
Fig. 7.
Fig. 8.
Fig. 9.
Fig. 10.
Fig. 11.
Fig. 12.
Fig. 13.
Fig. 14.
Fig. 15.
A
B
C

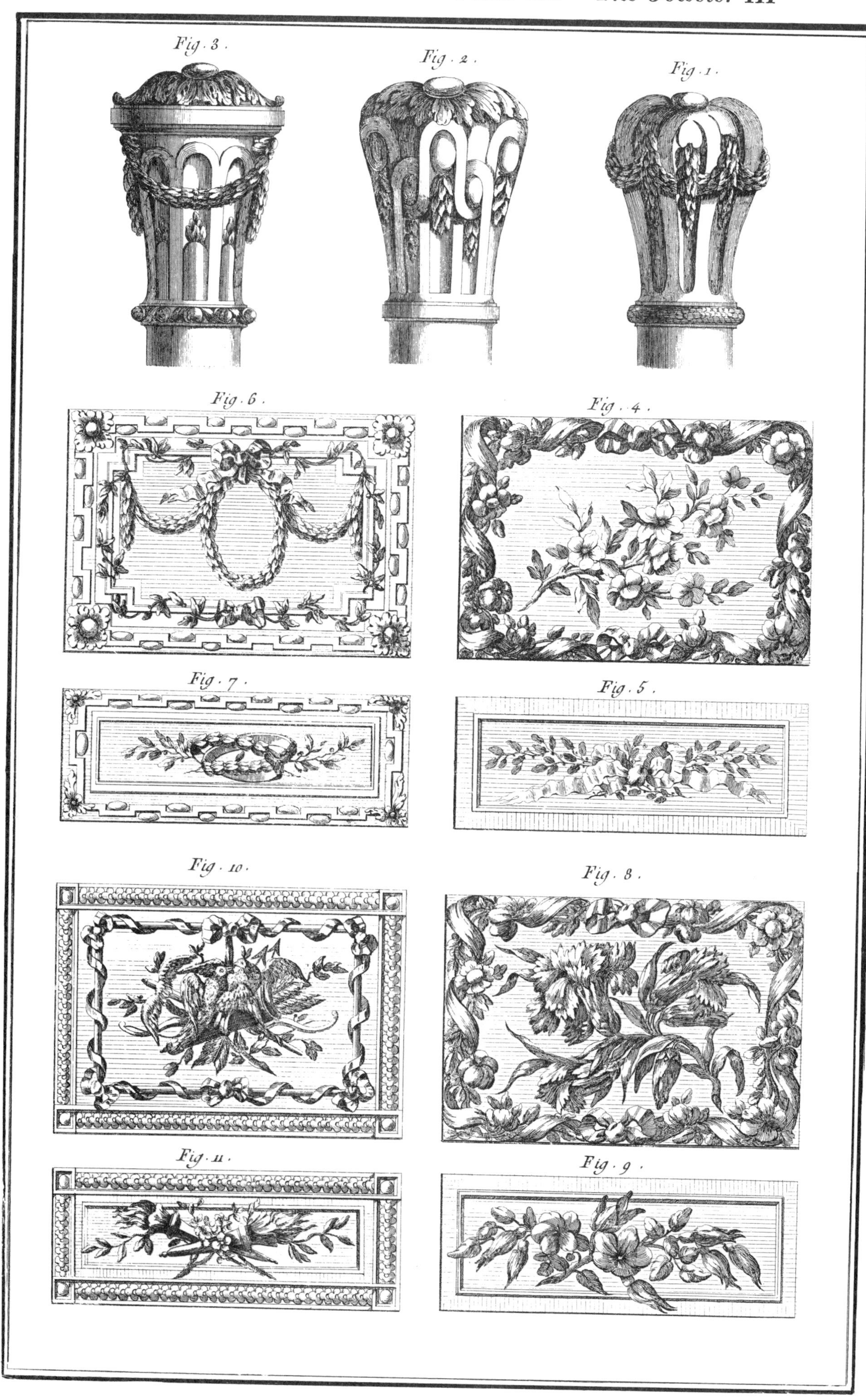
Fig. 3.
Fig. 2.
Fig. 1.
Fig. 6.
Fig. 4.
Fig. 7.
Fig. 5.
Fig. 10.
Fig. 8.
Fig. 11.
Fig. 9.

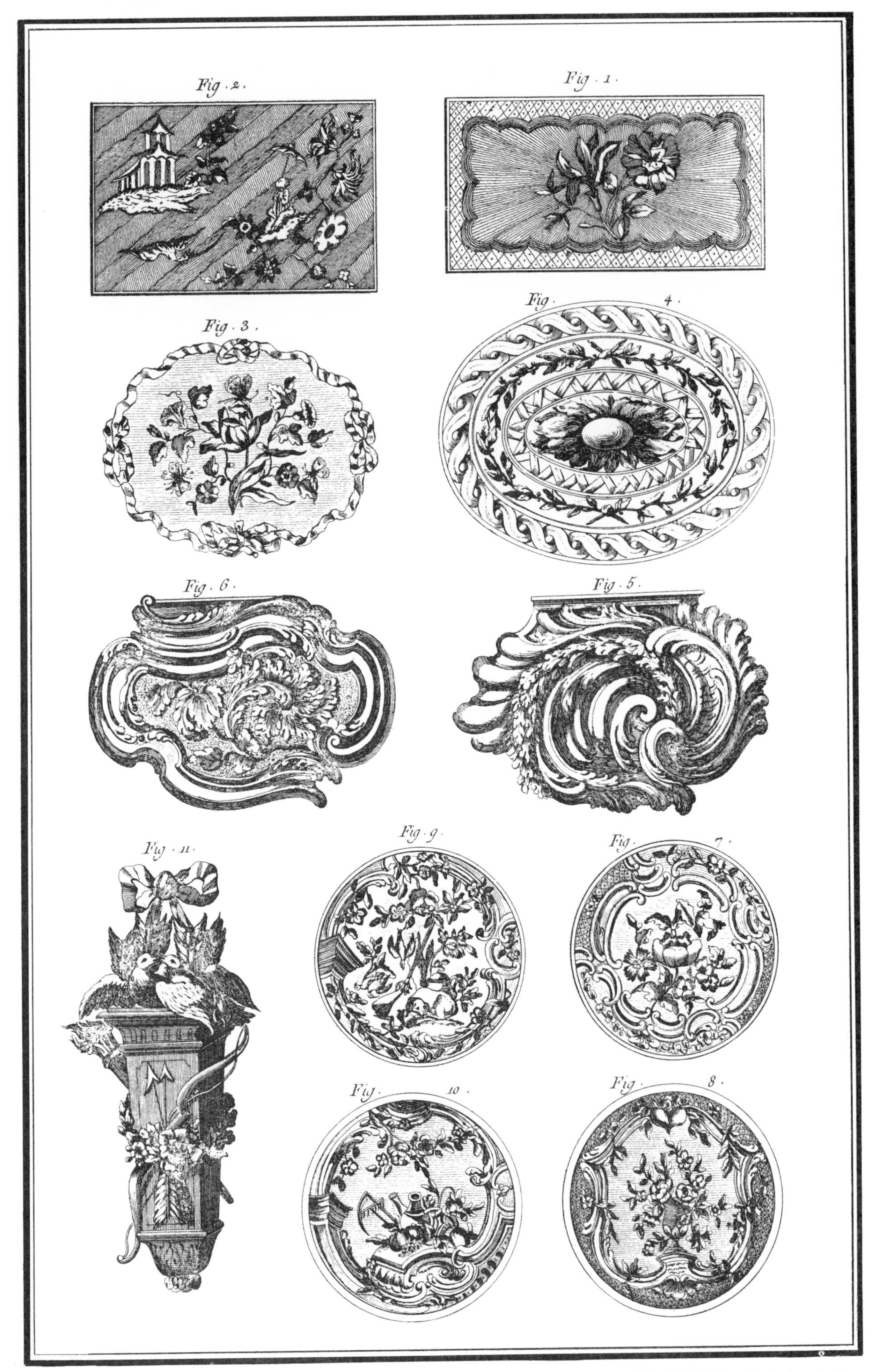

Vol. VIII, Orfèvre Bijoutier, Pl. IV.

Plates 412, 413, 414, 415, 416, 417
The Jeweler V, VI, VII, VIII, IX, X

Jewel-setters worked in a separate shop, back from the street. In this plate they are shown mounting diamonds.

The following plates illustrate settings—earrings, brooches, clips, and pendants—on which these craftsmen might have been working. The lower part of Plate 412 shows side, top, and bottom aspects of certain famous diamonds. Figs. 1, 2, 3 illustrate the Great Mogul, which weighed approximately 280 carats. Figs. 4, 5, 6 illustrate the Florentine or Tuscany diamond, then in the possession of the Grand Duke of Tuscany. It weighed 133 carats. Figs. 7, 8, 9, and 10, 11, 12 show the design of two brilliants owned by the King of France. The gem illustrated in 10, 11, 12, which weighed 547 grains, was reputed the most beautiful diamond in the world.

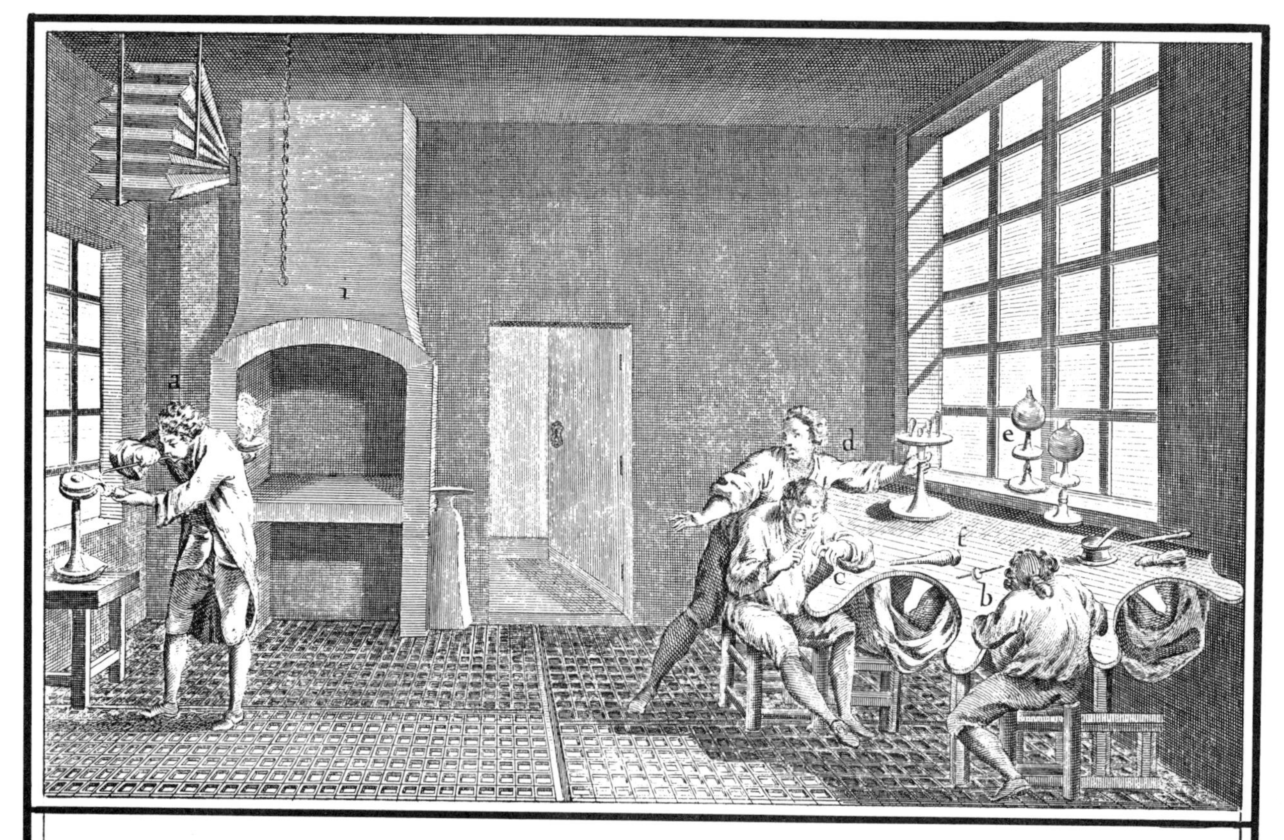

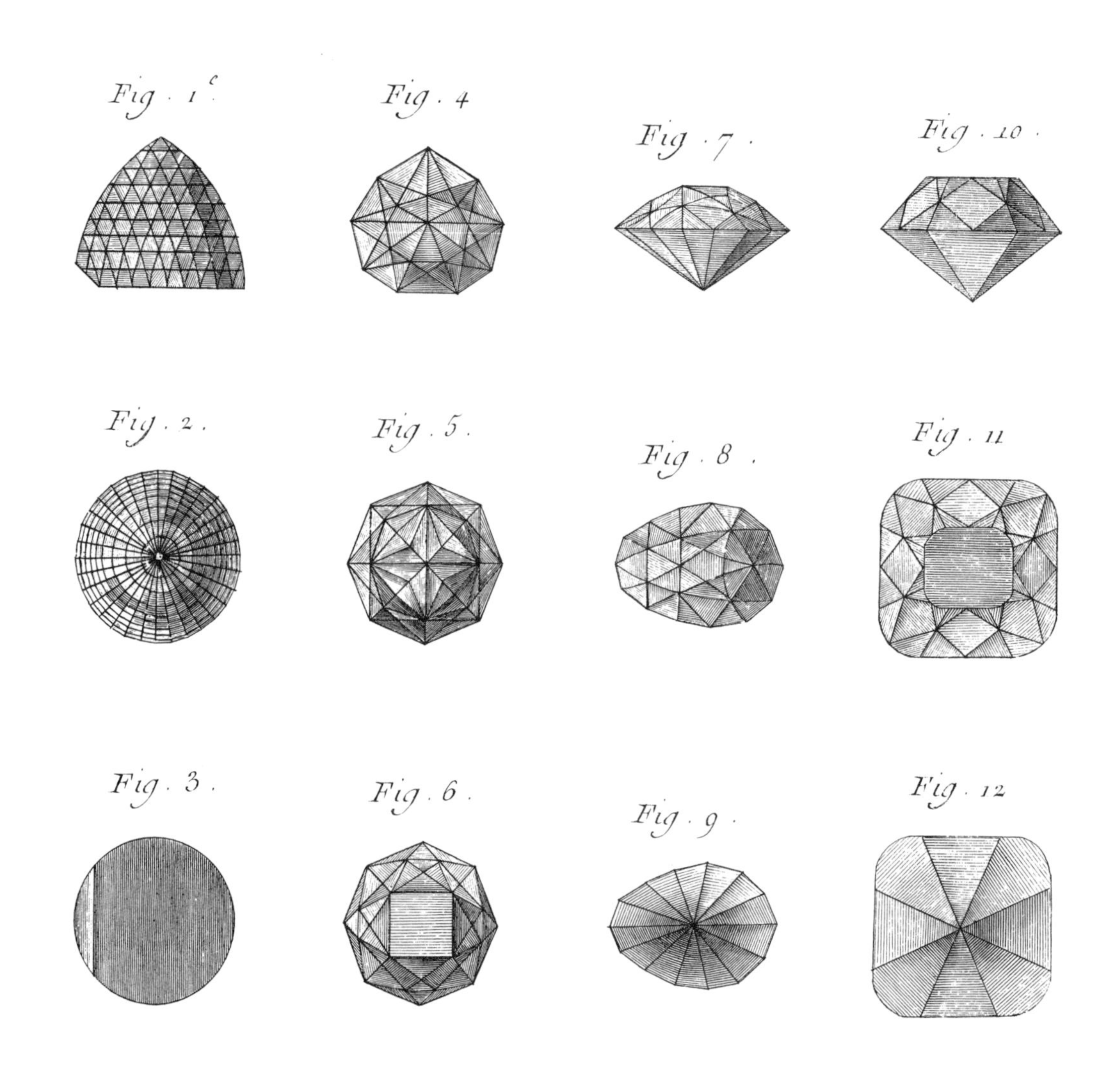

Vol. VIII, Orfèvre jouaillier, metteur en oeuvre, Pl. I.

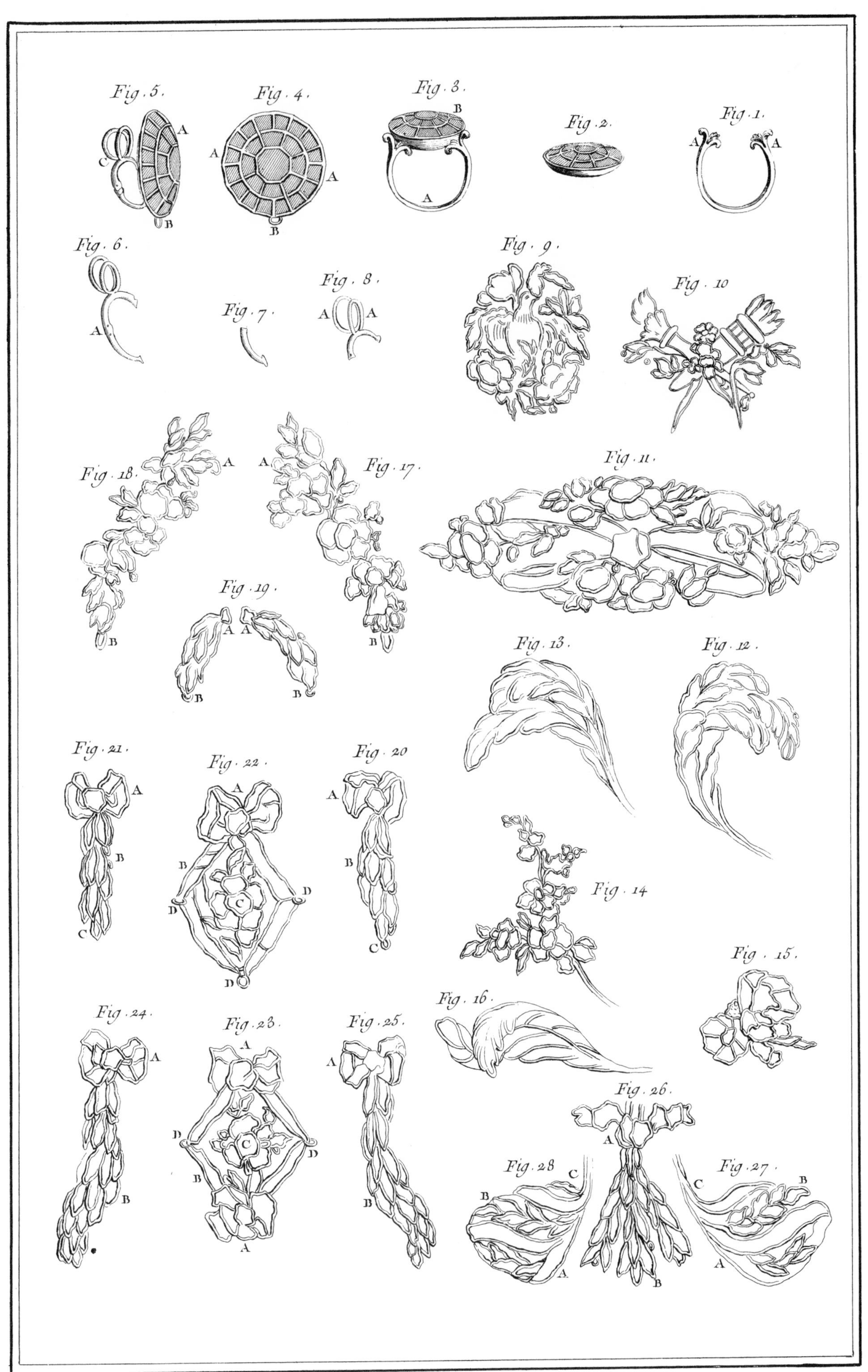

Vol. VIII, Orfèvre jouaillier, metteur en oeuvre, Pl. III.

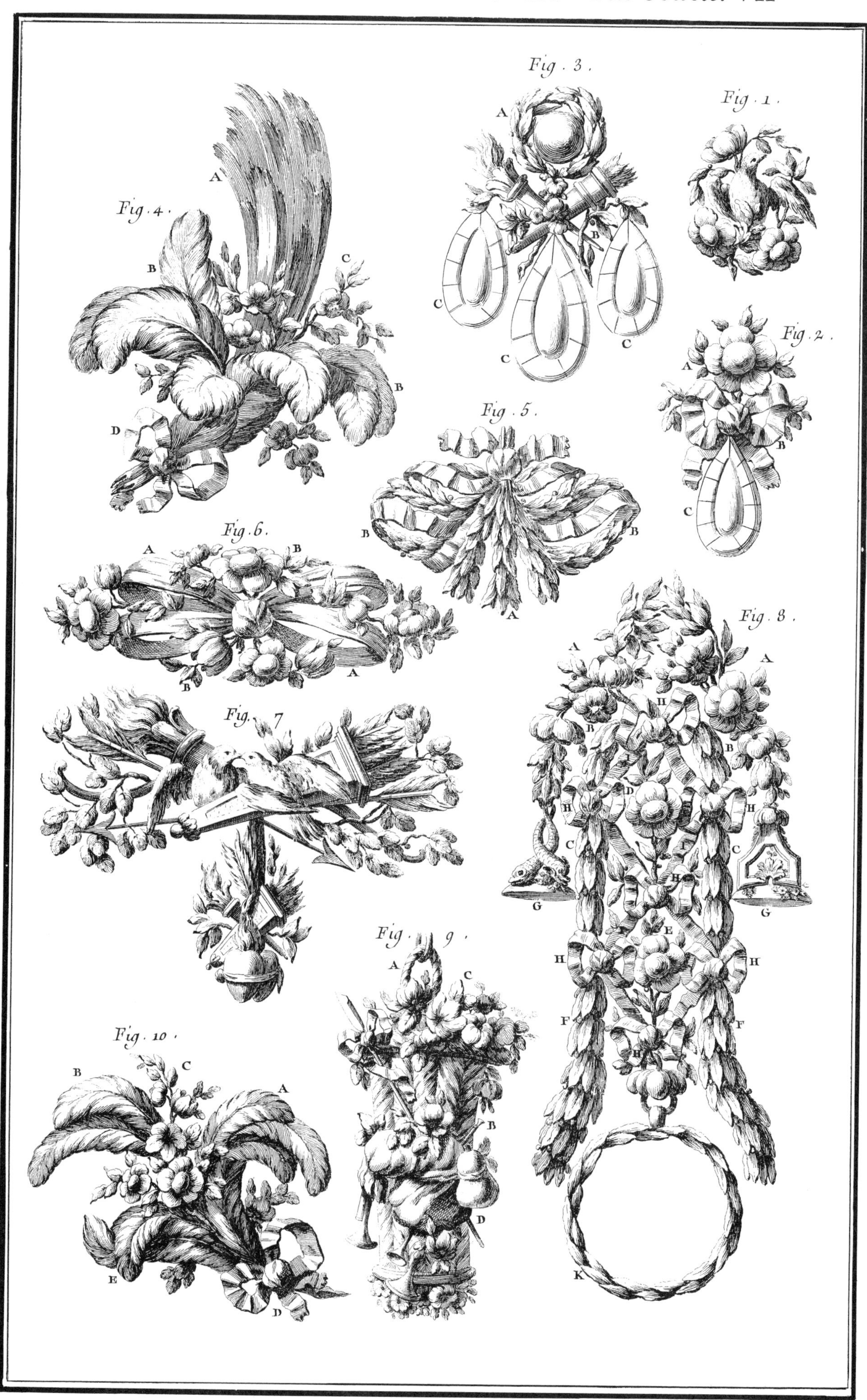
Fig. 1.
Fig. 2.
Fig. 3.
Fig. 4.
Fig. 5.
Fig. 6.
Fig. 7
Fig. 8.
Fig. 9.
Fig. 10.

Vol. VIII, Orfèvre jouaillier, metteur en oeuvre, Pl. V.

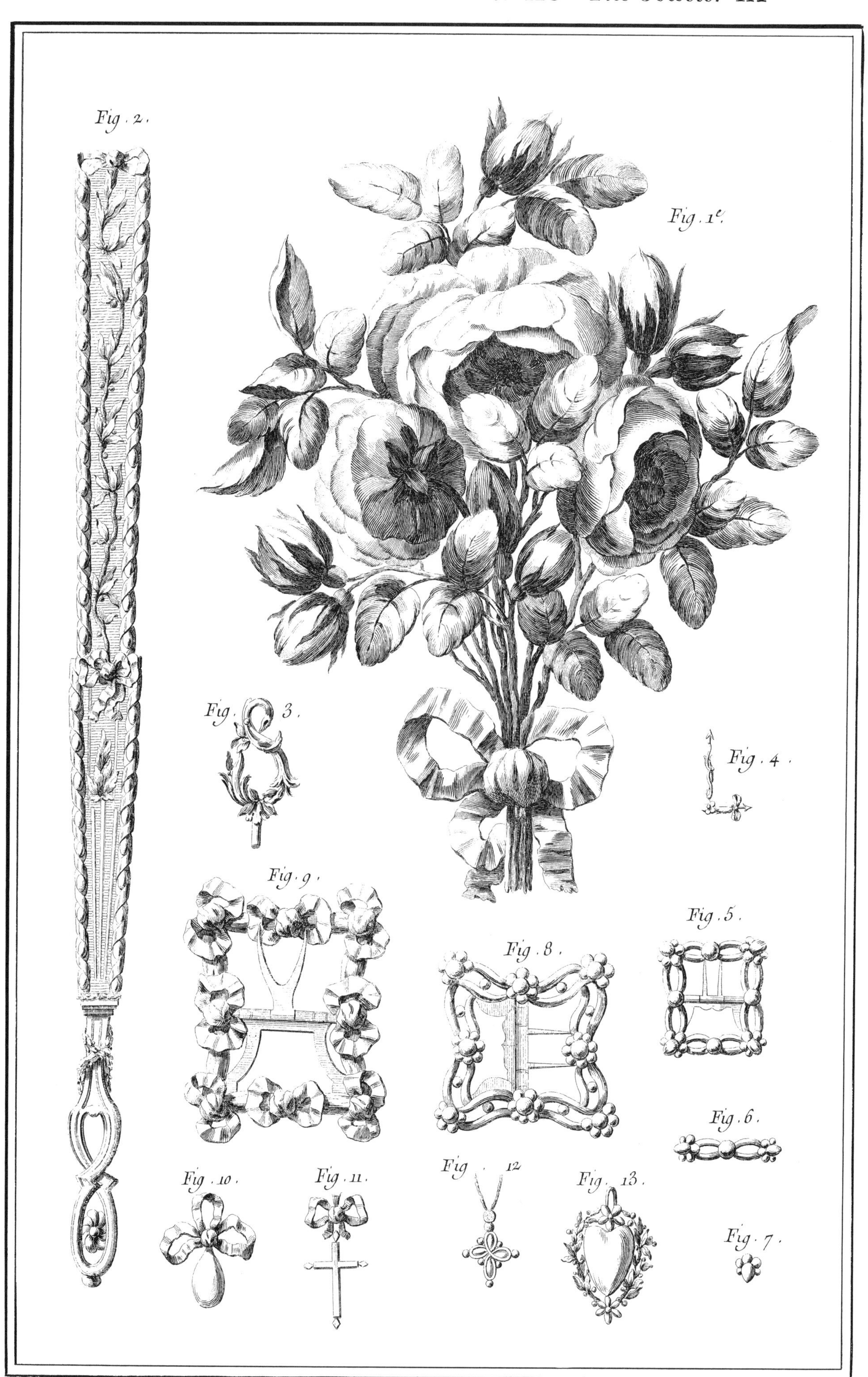

Vol. VIII, Orfèvre jouaillier, metteur en oeuvre, Pl. VI.

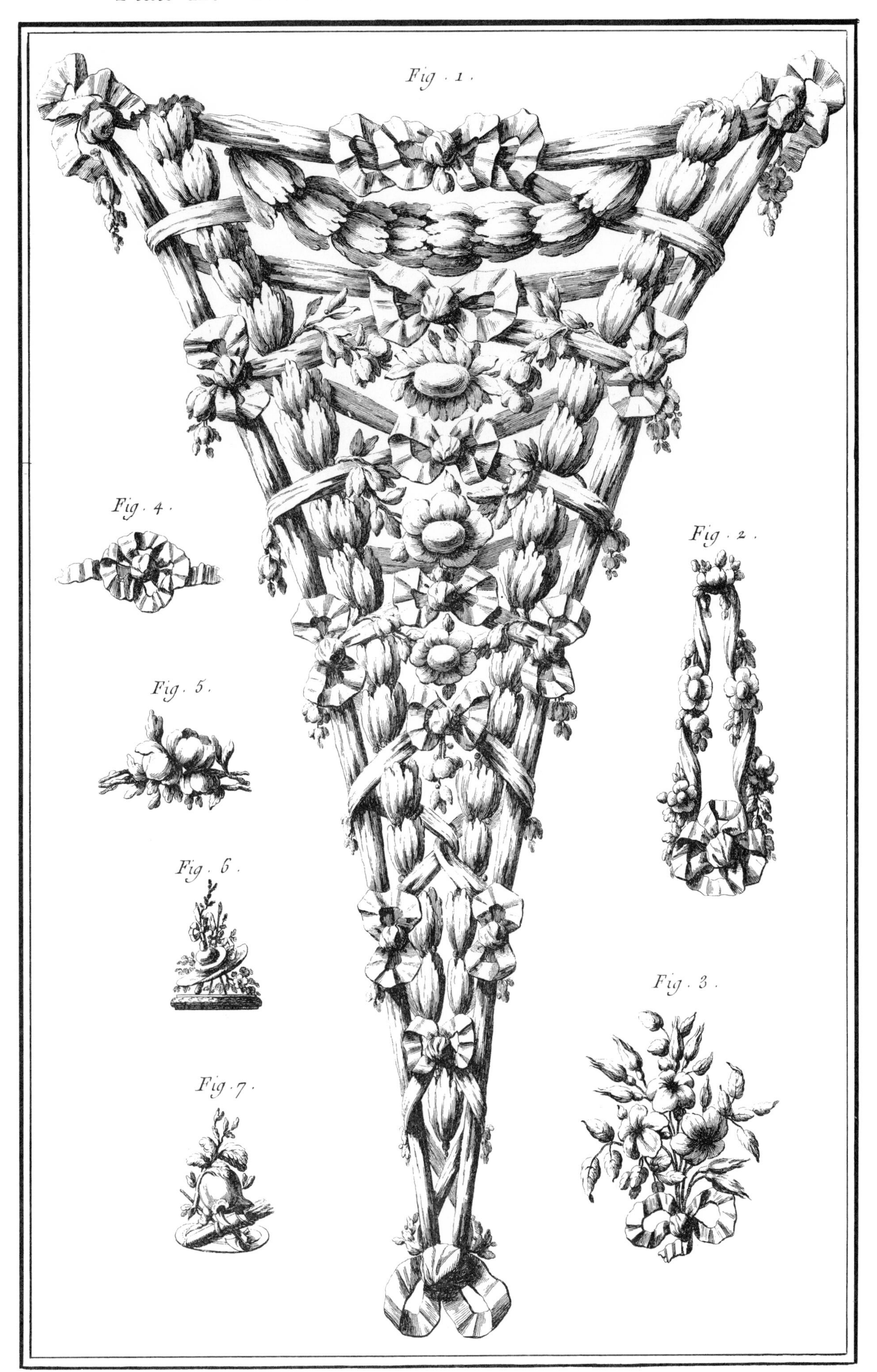

Vol. VIII, Orfèvre jouaillier, metteur en oeuvre, Pl. VII.

Plate 418 Goldbeating

The view of a goldbeater's shop gives very little impression of how lengthy and painstaking was the process of making gold leaf. It is one of the characteristics of gold that it can be beaten so thin, practically to a film, that it is possible to gild great domes without using an excessive amount.

As might be expected, the trade was a select one: the Parisian company of goldbeaters consisted of only 30 master-merchants. They bought their gold—only the finest quality would do—from the mint, and they liked to add to each melt (Fig. 1) a few ancient Spanish gold coins, doubloons or pieces of eight, since these were reputed to be of the purest gold ever mined. Borax was used as a flux, and ingots of bullion were cast in the little mold (a), preheated and lubricated with oil of palm.

The first step is to forge ingots into thin strips on the miniature (3 x 4 inch) anvil (b). These strips may then be further flattened either in a hand-rolling mill (Fig. 3, No. 2), or else by more beating—sometimes by both methods—by which time the thickness was about one-sixteenth of an inch. Now the gold is cut into strips of an inch by an inch and a half, and bound into laminated packets in which the layers of gold are separated by buffer strips of vellum. The structure of this pile was very complicated, requiring strips of parchment at intervals to lend the right resiliency.

It is this packet on which the beater of Fig. 2 is working, and will work for half an hour, tapping from the center outwards. Then the pile is disassembled (Fig. 3), the enlarged strips cut in half, and reassembled for a second beating. Even this is not the end, and yet a third hammering is required, in which the pile is laminated not with parchment and vellum, but with parchment and pieces of tanned skin taken from the abdomen of beef cattle. This particular skin, and no other, can be impregnated with calcined gypsum dust in such a way as to burnish the gold leaf in this, its final beating. As the Encyclopedist points out, here is an example of one of the secrets of craftsmanship, which at one time made the word "mystery" a synonym for a trade. How did men find out that this particular skin would impart the desired sheen to beaten gold? It was certainly not science that led them to the abdomen of a steer. Did they know about its properties when goldbeating began? Or did they find out about it through necessity?

Plate 418 _ Goldbeating

Vol. II, Batteur d'Or, Pl. I.

Plate 419 Silver Plating

Silvering represents the kind of craftsmanship for which France is famous. Utensils like those on the shelves and worktables are plated with silver, which is applied in layer after layer of paper-thin, finely-beaten silver leaf. The first and only unskilled step (entrusted to a woman, Fig. 1) is to scratch and cross-hatch copperplate so that the silver will have a rough surface to which to cling. In the case of articles to be chased or ridged in some way, the preparation is more difficult: then (Fig. 1, No. 2) the design must be chiseled into the base metal.

In the silvered key-hole opposite (Fig. 1, No. 3) is an example of how intricate the chiseler's work can be. He has mounted a piece on a block of cement for working. Before silver leaf is applied to metals, the surface must be blued on a charcoal brazier (Fig. 3 bis), and the vessel must be kept hot while the artisan applies silver leaf (Fig. 2), which he handles with calipers (Fig. 13) as delicate as a surgeon's, and presses on with a blunt tool (Fig. 8, No. 1) in his right hand.

Depending on the quality of the piece, it may be silvered to a depth of sixty leaves, after which it is burnished (Figs. 3 and 4) to the proper state of polish. Notice the artisans' air of sweet-tempered absorption in their tasks. Is the artist romanticizing old-fashioned craftsmanship? Or was it, perhaps, really good for the spirit?

Some idea of the quality of the work may be suggested by a glance at the tools: hatching knives (Fig. 8, No. 2; Fig. 11, No. 1 and 2; Figs. 12, 14) and polishers of all shapes (Figs. 6, 7, 9, 10).

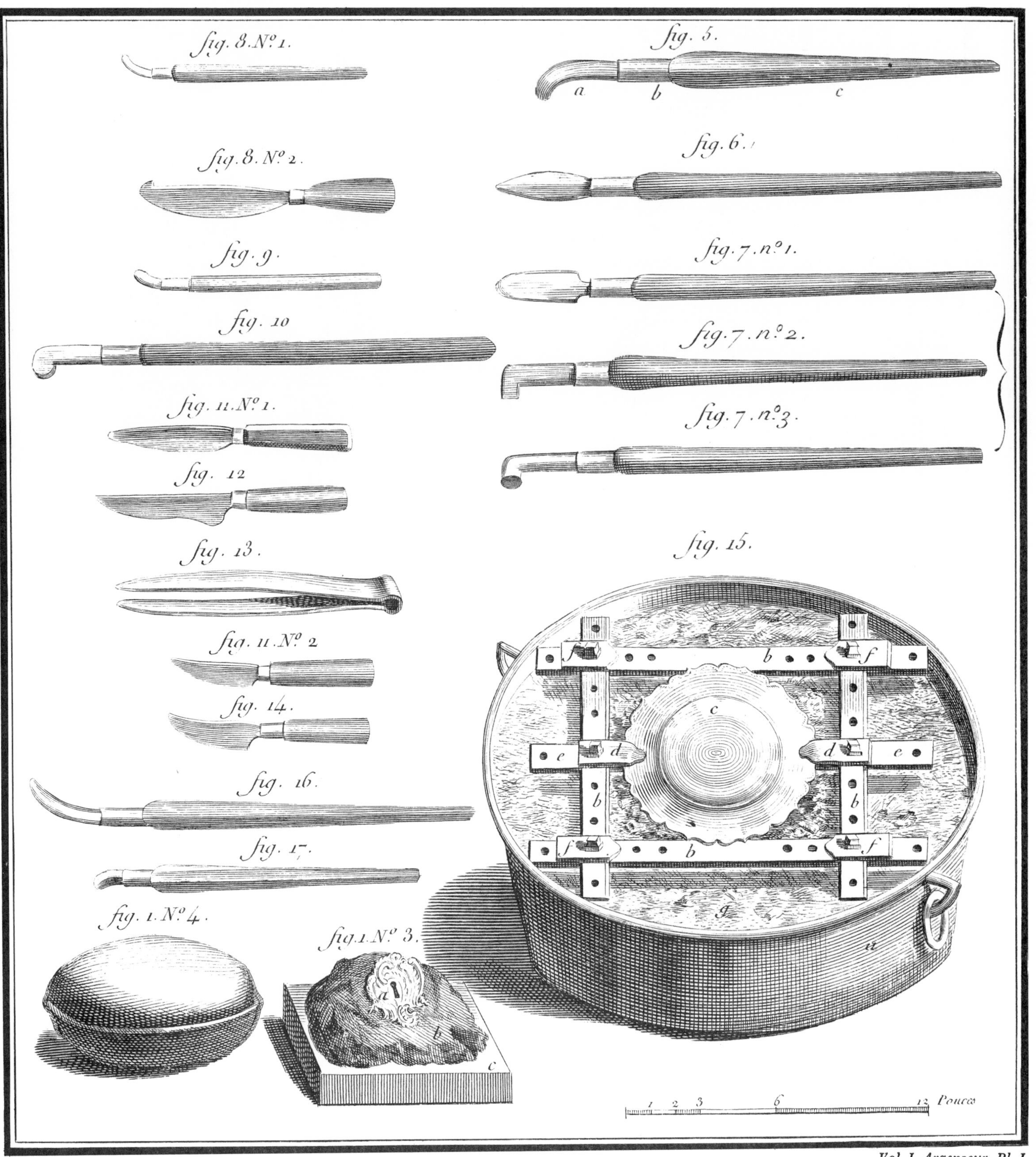
fig. 8. N.º 1.
fig. 5.
a
b
c
fig. 8. N.º 2.
fig. 6.
fig. 9.
fig. 7. n.º 1.
fig. 10
fig. 7. n.º 2.
fig. 11. N.º 1.
fig. 7. n.º 3.
fig. 12
fig. 13.
fig. 15.
fig. 11. N.º 2
fig. 14.
fig. 16.
fig. 17.
fig. 1. N.º 4.
fig. 1. N.º 3.
Pouces

Vol. III, Ciseleur Damasquineur, Pl.

Plate 420 Damascening

The art of chasing surfaces of steel with gold or silver patterns so as to make a metallic mosaic is known as "damascening," after the city of Damascus, whence it had been imported to Italy in medieval times and from there to France in the 16th century. In this its history is typical of many of the arts requiring exquisite taste and workmanship which have distinguished French enterprise ever since it experienced the stimulating and refining influence of Renaissance craftsmanship.

As may be seen from the objects at the back of the shop—a sconce, a sword hilt, a candlestick, fine muskets—damascening lends an ultimate luxury to the ornamentation of metalwork. The process, as the Encyclopedia *points out, is a good deal easier to understand than to practise: it consists simply of etching and chiseling the design (Figs. 1, 2), applying threaded gold (Figs. 3, 4), tempering gently over a brazier (Fig. 5), burnishing (Fig. 6), and polishing (Fig. 7).*

Vol. III, Doreur, sur métaux, Pl. I.

Plate 421 Gilding I

The taste of the 18th century inclined (perhaps too far) in the direction of ornamentation and magnificence. It might truly be called a gilded age, for gilt was applied very lavishly to any surface which would take it. In the case of this shop, metals are being gilded. The operations at the right prepare the metal for its golden coat: gentle heating (Fig. 1); dilute acid for cleaning (Fig. 2); blueing (Fig. 9). Gilt itself is brushed on (Figs. 6 and 7) and the object gently buffed (Fig. 8).

Plate 422 Gilding II

To gild leather it is first necessary to print, or emboss, the design on the leather piece. This is accomplished by wood blocks (Fig. 13, opposite). Leather would usually be embellished in various colours with paints (Fig. 1), as well as gold or silver leaf, or both. In either case the metal is applied with a pair of ebony pincers (Fig. 2) on the handle of which is mounted a piece of foxtail which brushes without scratching. Finally, the gilding is pressed into the bosses (Fig. 5), and a polisher gently buffs a gilded leather band (Fig. 4). The tools appear opposite.

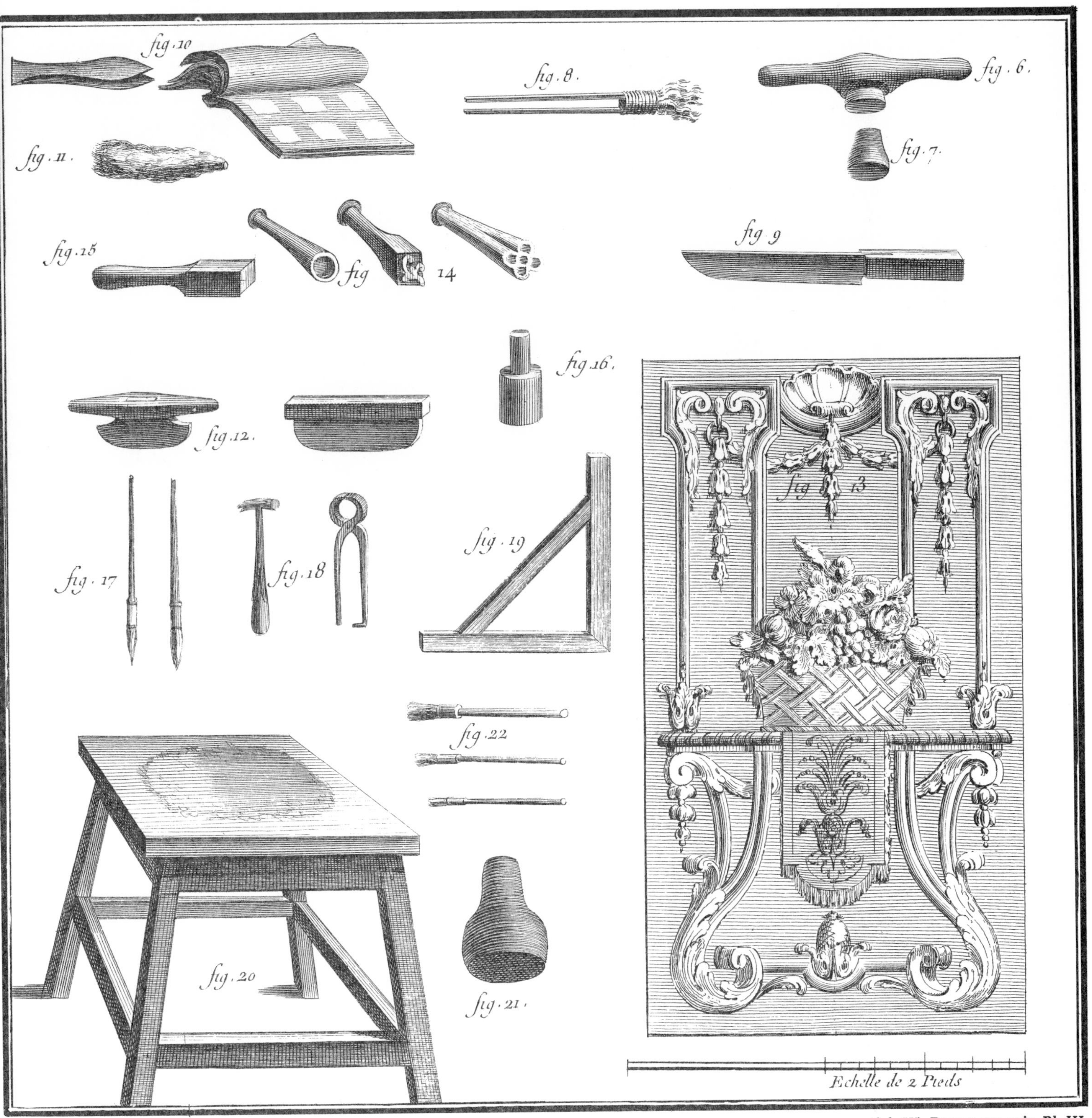
fig. 10
fig. 8.
fig. 6.
fig. 11.
fig. 7.
fig. 15
fig
14
fig 9
fig. 16.
fig. 12.
fig 13
fig. 19
fig. 17
fig. 18
fig. 22
fig. 20
fig. 21.
Echelle de 2 Pieds

Plate 423 Gilding III

The supplementary volume of plates illustrates a new method of gilding or silvering leather, one that was strongly advocated by Fougeroux de Bondaroy, one of the members of the Royal Academy of Science who interested himself particularly in technological matters.

In the picture above hides are curried (Figs. 1-3), cut to size (Fig. 4) and hung to dry (Fig. 5). In the outdoor enclosure opposite, the hides are varnished (Fig. 6) and passed on, first to the spreader (Fig. 7) who evens out the varnish, and secondly to a beater (Fig. 8) who roughs up the surface by tapping it with his fingertips, which makes it take gilding more firmly. On the other bench, the design to be gilded is etched into the varnished surface with a knife, and the corners and edges wiped clean (Fig. 10). The actual gilding is pressed in as in the old method (see Plate 422), but for the rigid upper plate Fougeroux would substitute the assembly (GHIK right). I fits into K and between these and K are slipped several thicknesses of cardboard to cushion the press action and avoid cutting around the edges of the relief pattern.

Supplement, Doreur sur cuir.

Vol. III, Doreur, sur bois, Pl. I

Plate 424 Gilding IV

Picture frames illustrate the gilding of wood. The surface has to be primed (Figs. 2 and 5) with a sort of paste, before the gilt is applied (Fig. 3).

Vol. IV, Emailleur, à la lampe, Pl. I.

Plate 425 Enamelling I

Through the strands of French civilization runs a welcome strain of femininity emerging in sweetness of taste and delicacy of perception. An example is the art of enamelling, which brought out the best in French craftsmanship. The technique is that of a small, refined glass working. In essence enamels are glassed surfaces (usually metallic) which owe the distinctive lustre and permanence of their colors to vitrification.

In the 17th and 18th centuries French enamels were works of fine art. As the Encyclopedist writes enthusiastically: "Of all the arts I know it is the most agreeable and the most amusing. There is no object which one cannot execute in enamel by the heat of the lamp, and that in very little time." And he urges it on his readers as a hobby, even though in connection with the last of these plates he feels it necessary to warn them, "If it is true in all the arts that the distance from the mediocre to the good is great, and that from the good to the excellent almost infinite, this is a truth particularly striking in the painting of enamels."

Vol. IV, Peinture en Email, Pl. I.

Plate 426 Enamelling II

The most delicate and characteristically 18th century enamels, the medallions or lockets bearing portraits or exquisite small designs, were prepared as miniature paintings (Fig. 1) on a surface enamelled in neutral shades. The color would be glazed into the enamel in a small but very hot oven (Fig, 2). The artist's task was doubly difficult. Not only was his work on a tiny scale, but glazing altered the colors. He had to imagine therefore, or rather to know, what his colors would be when they emerged from firing, for they would not be as he painted them.

Vol. IV, Emailleur, Pl. II.

Plate 427 Artificial Pearls I

Akin to enamelling was the manufacture of artificial pearls. They are made from the tube of opalescent glass which the worker in Fig. 1 is about to cut into small tubes for blowing. At the left-hand table, a girl (Fig. 2) blows a pearl and her companion (Fig. 3) smooths the edges where it is broken from the tube. The other two (Figs. 4 and 5) seem to be engaged in a different operation. They are spinning glass thread (4) and winding it onto a reel (5).

Plate 428 Artificial Pearls II

Artificial pearls were colored by an extract drawn from the scales of a small, fresh-water fish, the ablette, *or bleak, one of which the girl of Fig. 1 is cleaning. After soaking in hot water for half a day, the scales yield a pearly coloring matter,* essence d'Orient *which a second girl (Fig. 2) sucks into her pipette. A single drop of this liquor is blown into each pearl (Fig. 3), which is then graded for size. At the center table the pearls are filled in a basin of melted wax, and finally the hole for the string is opened by passing a tight roll of strong paper through each pearl. Details of the techniques appear opposite: the grader (Fig. 10), the paper rolls (Figs. 4-8), the device for letting the pearls into the wax (Fig. 9).*

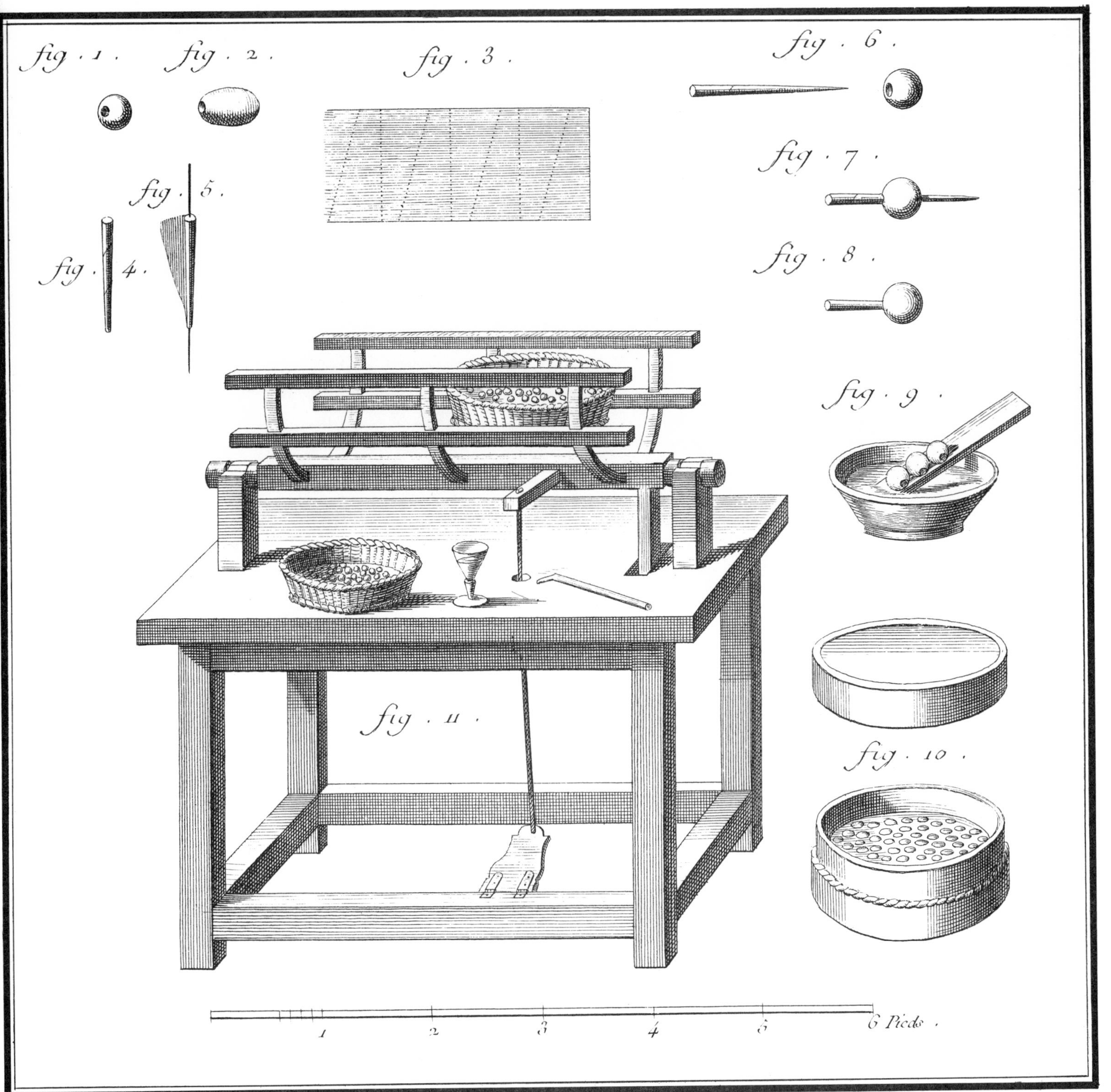
fig. 1.
fig. 2.
fig. 3.
fig. 4.
fig. 5.
fig. 6.
fig. 7.
fig. 8.
fig. 9.
fig. 10.
fig. 11.
1
2
3
4
5
6 Pieds.

Fashion

Supplement, Marchande de Modes.

Plate 429 Haute Couture

"Surely," exclaimed Voltaire in a moment of exasperation with French frippery and flippancy, "we are the whipped cream of Europe." But he was a man who appreciated style, and it is doubtful that he really meant what he said. In any case, the French have in practice refuted that Puritan or Spartan economic prejudice which represents luxury as impoverishing.

This is the establishment of some 18th century Dior, for Paris has always known how to make fashion pay, and men as well as women had their fashions.

Plate 430 Feathers I

So dependent was the milliner on feathers, so nice the distinctions at court expressed by the length or the curl of a plume, that the plumassier's *was a full-fledged trade. Her ceiling hung with peacock tails and ostrich plumes, Madame la plumassière (a) is at work upon the toque surmounting a headdress prescribed for a peer of France, this one a duke, at official occasions.*

Plate 430 Feathers I

Vol. VIII, Plumassier-Panachier, Pl. I.

Theatre-goers who had the good fortune to see the Madeleine Renaud and Jean-Louis Barrault performance of Le Misanthrope *in that company's New York visit in 1957 will appreciate how essential was some such hat to a young gentleman about town, not so much for covering his head, as for flourishing in his hand.*

The elaborate construction in the center (c) is also destined for a ceremonial function. It is to be worn by a horse which pulls the carriage of an ambassador to the King of France to a reception. At the right the lady's gown is tricked out in ostrich bows.

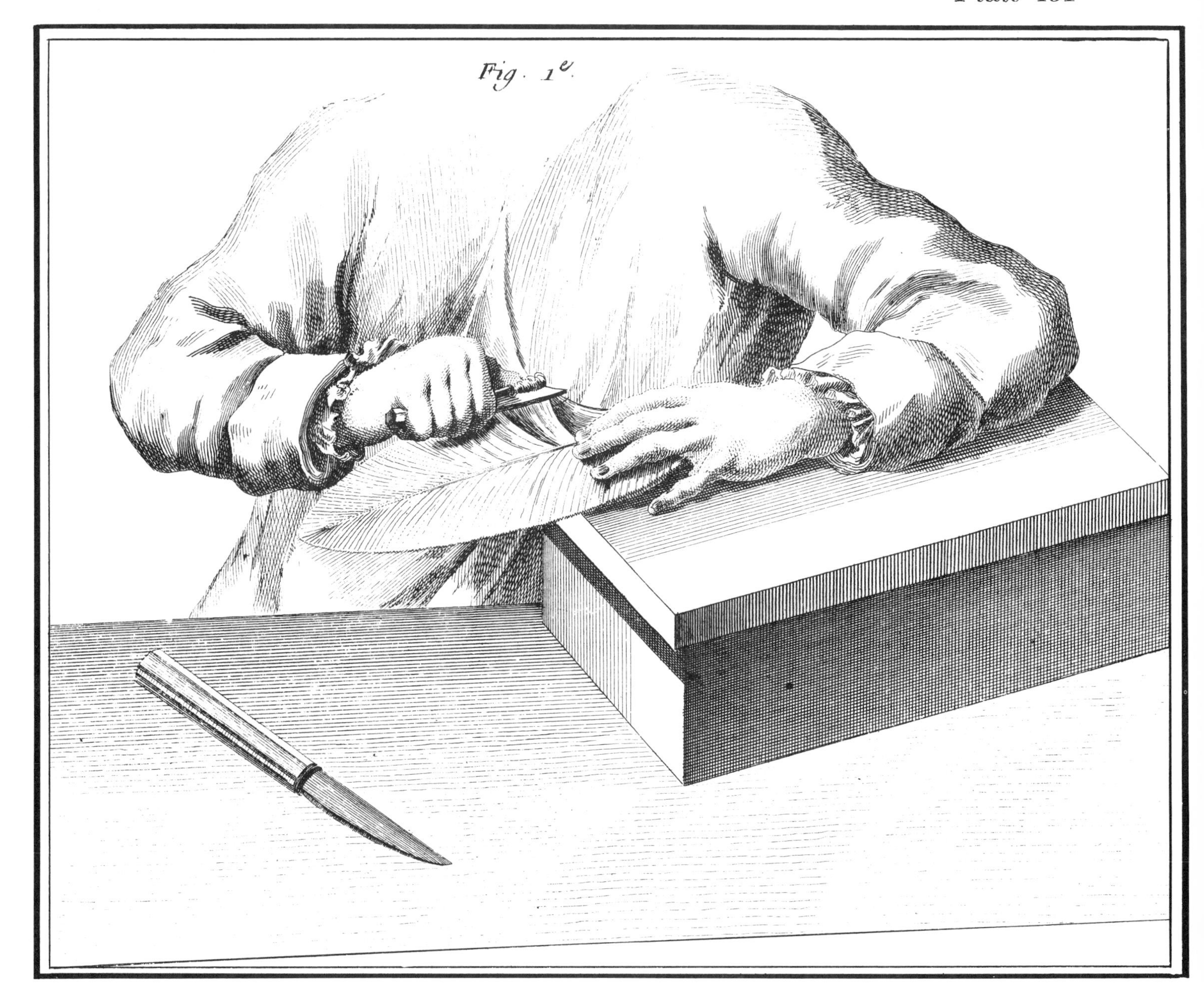

Plate 431 Feathers

The ostrich feather in a state of nature was too heavy for the world of fashion, and too plain: it had to be thinned (Fig. 2) and frizzed (Fig. 1). Opposite (Fig. 5), is the rarest and most expensive of feathers, the heron's plume. Of the two creations, the lower (Fig. 4) is a nobleman's panache, *and the more chaste bonnet of Fig. 3 would be worn by novices at their reception into the order of the Holy Spirit.*

Vol. VIII, Plumassier-Panachier, Pl. II.

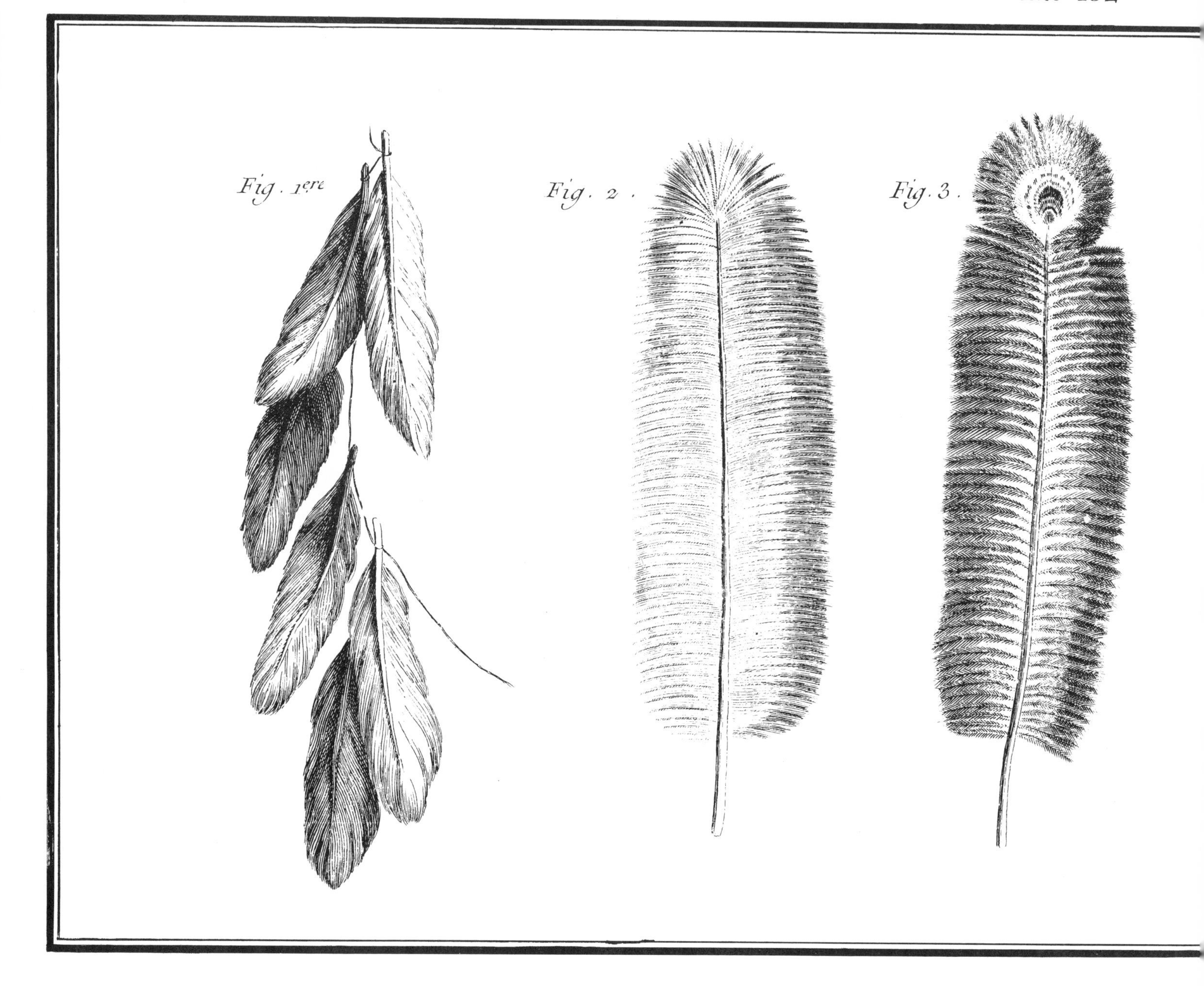

Plate 432 Feathers III

Various feathers before curing: ostrich (Figs. 1, 2, 4, 5); peacock (3); cock (6); swan (9). Figs. 7 & 8 are hat-feathers all treated and ready to stick in the band.

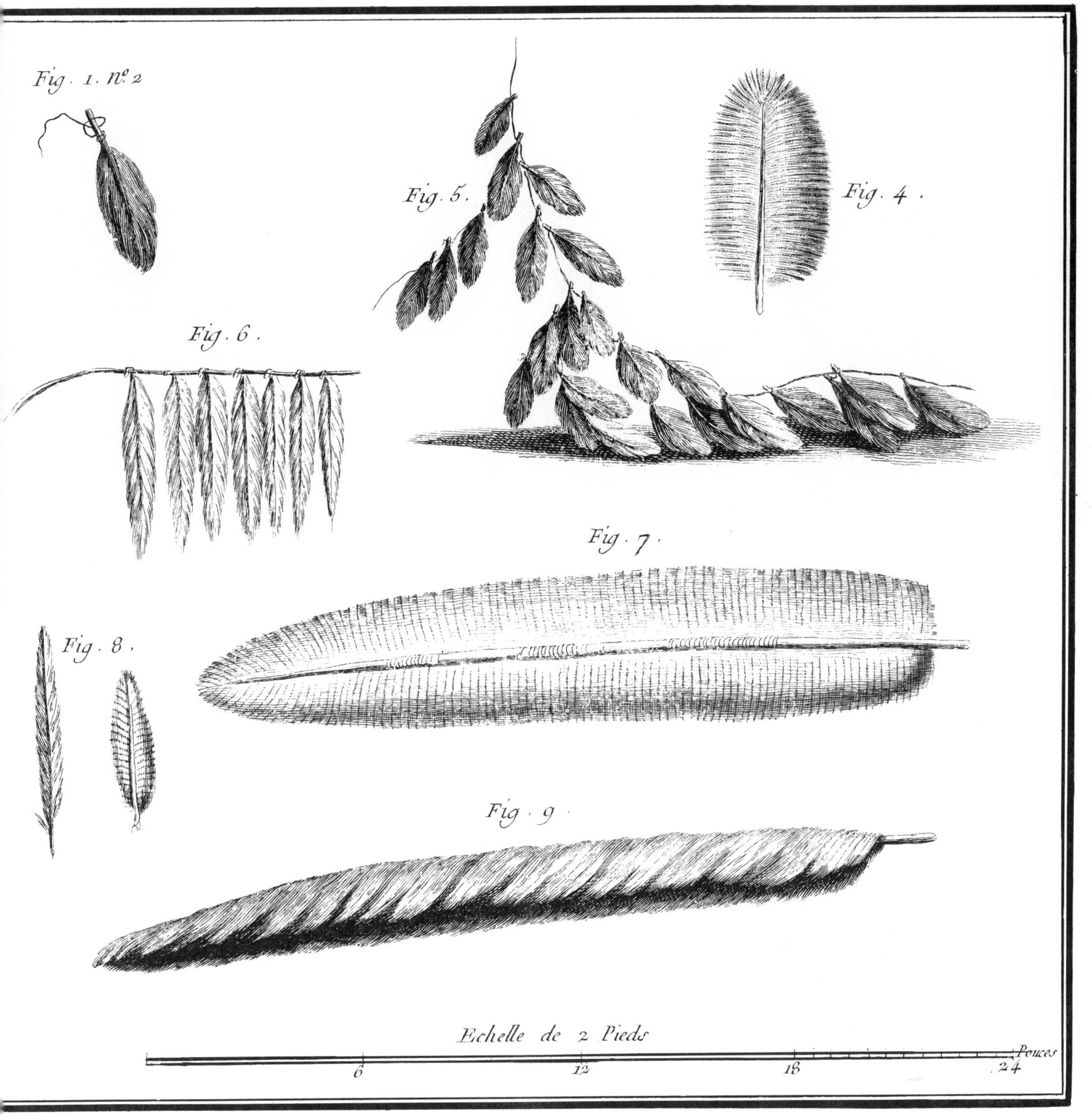
Fig. 1. No. 2
Fig. 5.
Fig. 4.
Fig. 6.
Fig. 7.
Fig. 8.
Fig. 9.
Echelle de 2 Pieds
Pouces
6
12
18
24

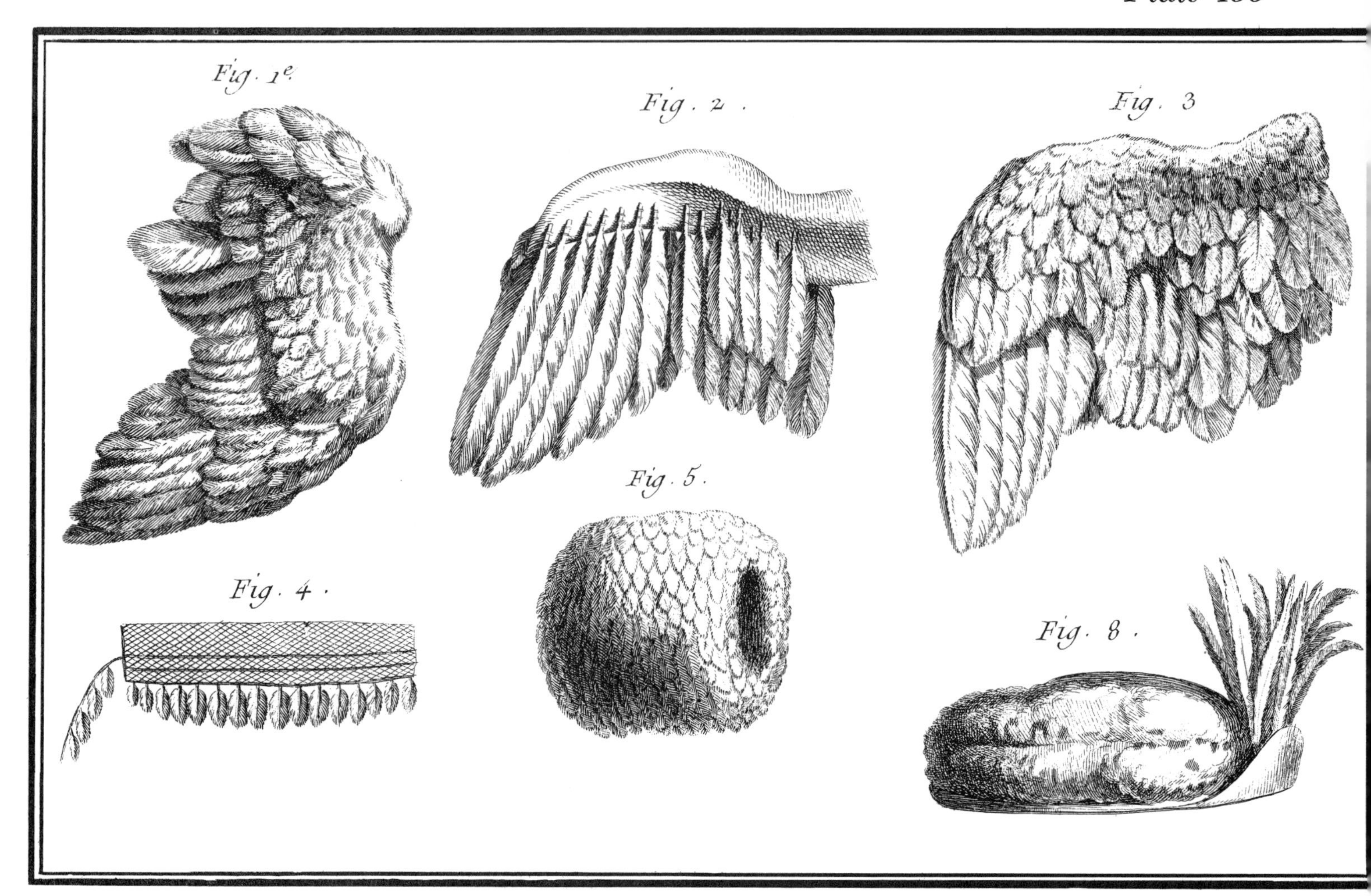

Plate 433 Feathers IV

Of these creations of the plumassier, *the wings (Figs. 1-3) are to be worn by children taking the parts of angels in religious processions. Below are a pair of muffs, of small, undefined feathers (5) and cock plumes (7), each shown with its starting design at the left (4, 6). The smartly restrained bonnet of Fig. 8 bears the heron plumes of Fig. 14. Fig. 9 is a lady's hat in the Indian style—the American Indian enjoyed a great vogue in 18th century Paris—while the small plumes of Figs. 15-17 were thought suitable for a child's bonnet where ostentation would be out of place. Helmets were not seen in Paris—that of Fig. 10 is what an actor would wear portraying a king at the Comédie Française. Finally, of the three* panaches *at the bottom, the most ornate (Fig. 12) was made to surmount the pole supporting a canopy at some courtly function, whereas the one of Fig. 11 would bob up and down on the head of a horse, and that of Fig. 13 between the ears of an aristocratic mule.*

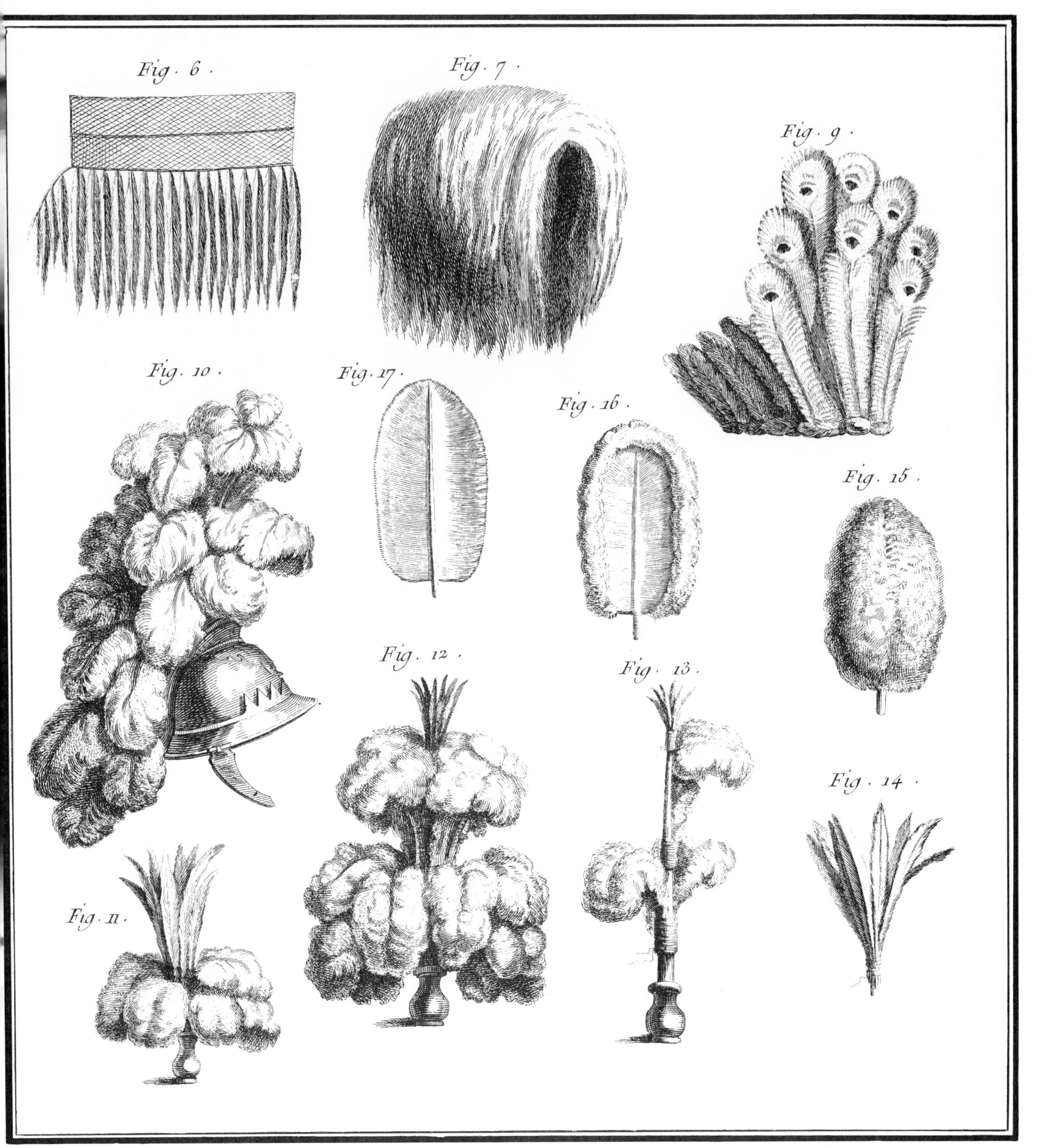
Fig. 6.
Fig. 7.
Fig. 9.
Fig. 10.
Fig. 17.
Fig. 16.
Fig. 15.
Fig. 12.
Fig. 13.
Fig. 14.
Fig. 11.

Vol. IV, Fourreur, Pl. V

Plate 434 The Furrier

In the 18th century men as well as women formed the furrier's clientele. French shop-keeping is remarkable for extreme specialization, and it would appear that this shop stocked mainly muffs, all distributed around the walls according to size, although there is one fur-trimmed skirt out for display (f). The establishment makes its own muffs from skins (c) which it buys from wholesale dealers. The elegance of the shop befits the business, and so does its comfort. This is no store-front open to the weather. The windows are glazed, and the customers warmed by a stove (h) as discreetly decorated as a commode.

Plate 435 The Wigmaker I

Vol. VIII, Perruquier-Barbier, Pl. I.

Plates 435, 436, 437 The Wigmaker I, II, III

Wigs went out of fashion with the 18th century, their decline hastened, no doubt, by French revolutionary ideals of equality and Republican simplicity in dress. Wigs were an aristocratic item of dress, and though the great majority of the aristocracy survived the Revolution physically, as a class it never regained its composure. The wigmaker followed his clientele to oblivion. Men might still be shaved (a) in the 19th century, but it was at home or at the barber's. Curling irons might still be heated (e), and surplus powder wiped off the face (f), but it was not for men that these services were performed. The creations illustrated in the following plates are those of an extinct art.

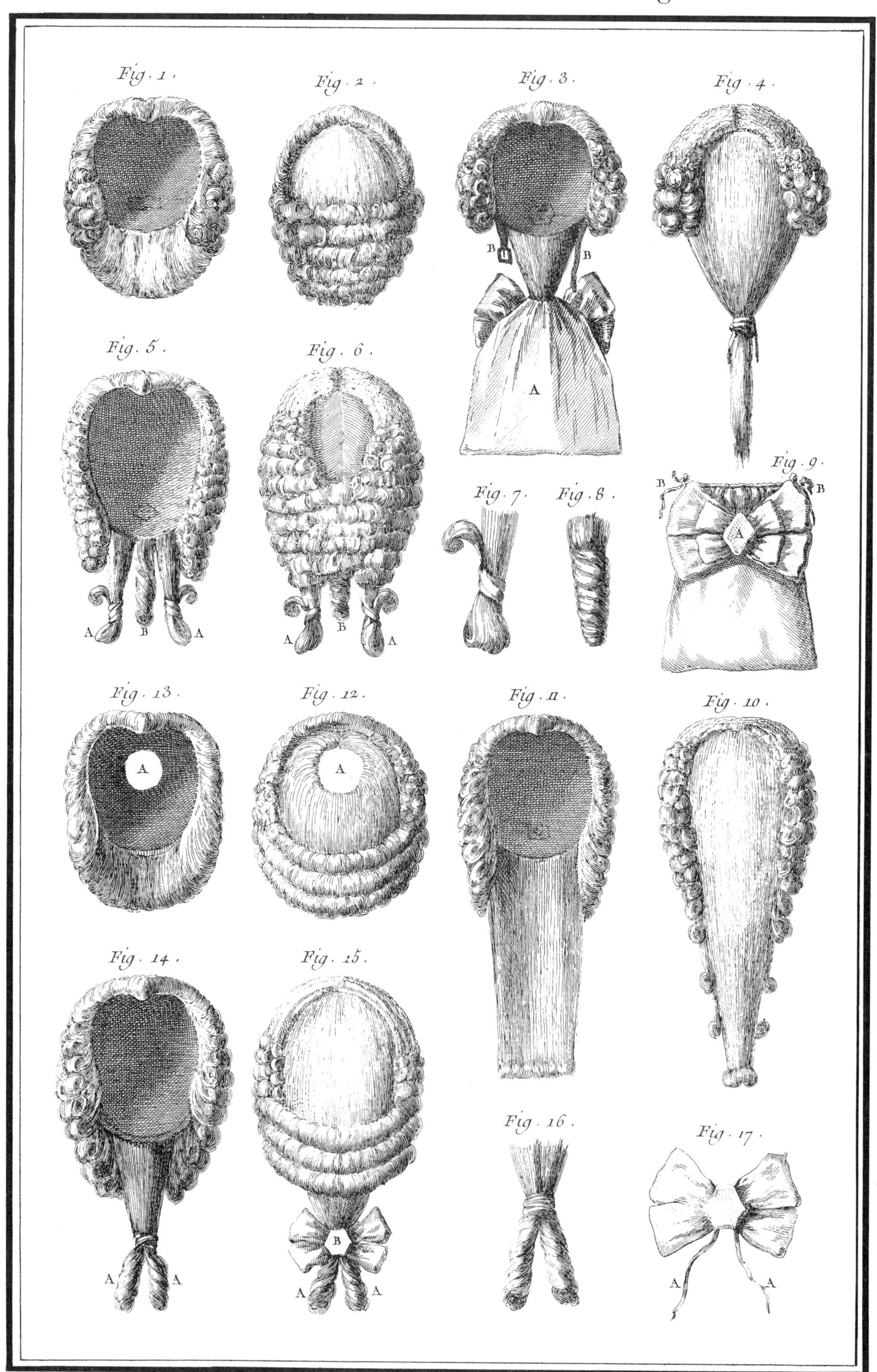
Fig. 1.
Fig. 2.
Fig. 3.
Fig. 4.
B
B
A
Fig. 5.
Fig. 6.
Fig. 7.
Fig. 8.
Fig. 9.
B
B
A
A
B
A
A
B
A
Fig. 13.
Fig. 12.
Fig. 11.
Fig. 10.
A
A
Fig. 14.
Fig. 15.
Fig. 16.
Fig. 17.
B
A
A
A
A
A
A

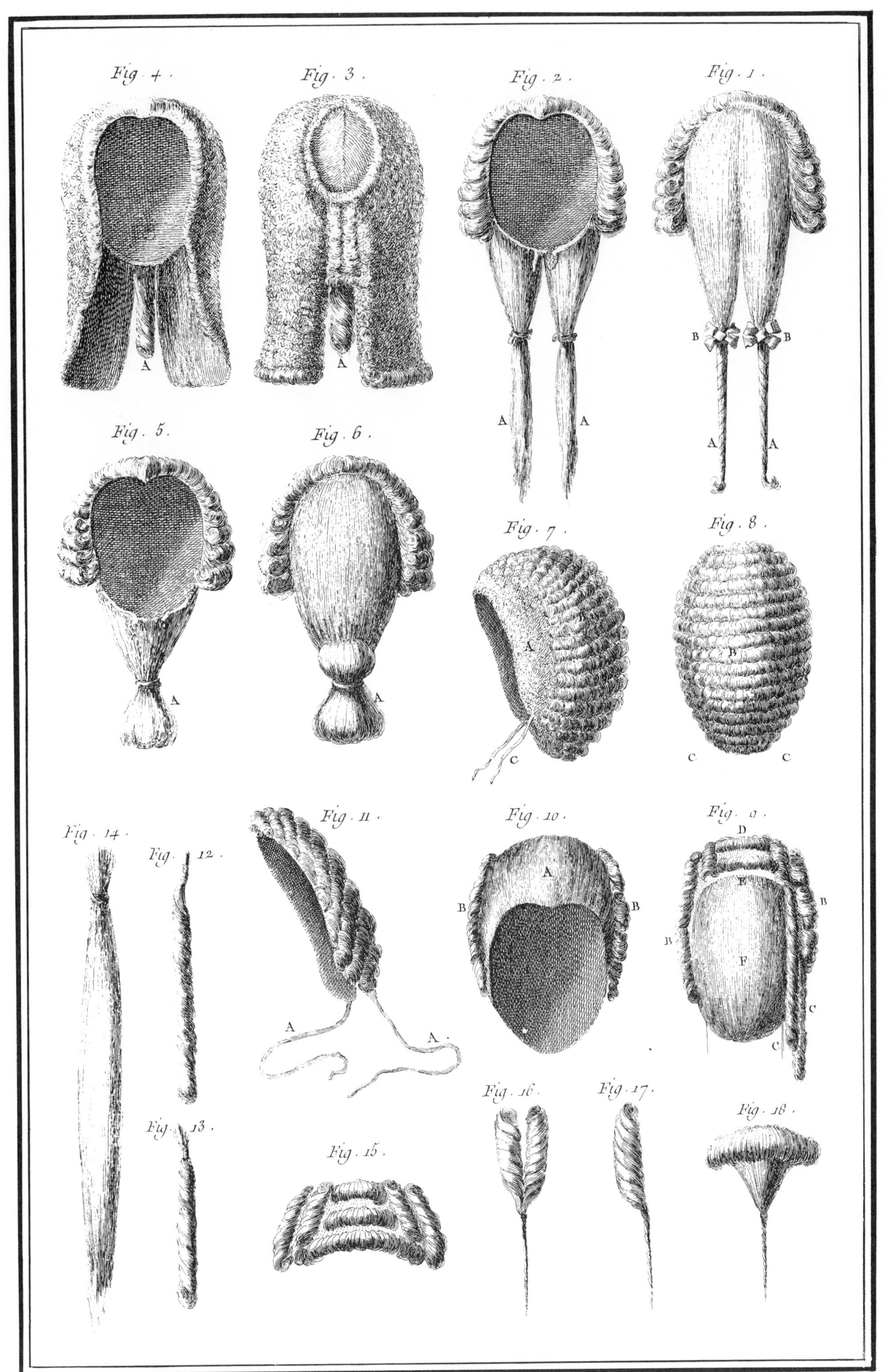
Fig. 4.
A
Fig. 3.
A
Fig. 2.
A
A
Fig. 1.
B
B
A
A
Fig. 5.
A
Fig. 6.
A
Fig. 7.
A
C
Fig. 8.
B
C
C
Fig. 14.
Fig. 12.
Fig. 13.
Fig. 11.
A
A.
Fig. 10.
A
B
B
Fig. 9.
D
E
B
B
F
C
C
Fig. 15.
Fig. 16.
Fig. 17.
Fig. 18.

Plate 438 The Pursemaker I

Vol. I, Boursier, Pl

Plates 438, 439 The Pursemaker I & II

The rather drab interior of this pursemaker's shop gives no hint of the variety of handbags and other items which emerge from the hands of the two apprentices and their master, who has stepped out. In the 18th century gentlemen as well as ladies carried handbags, and their taste was equally capricious.

The pursemakers also had the right to make leather breeches and certain types of caps —among them, oddly enough, clergymen's birettas and hunting caps—and parasols, for which the boy at the bench (Fig. 1) is cutting a length of brass wire. The range of the pursemaker's wares is illustrated in the next plate. The device at the top (Figs. 1 and 2), though it resembles a birdcage, is in fact a lantern; its shutter opens along the dotted line.

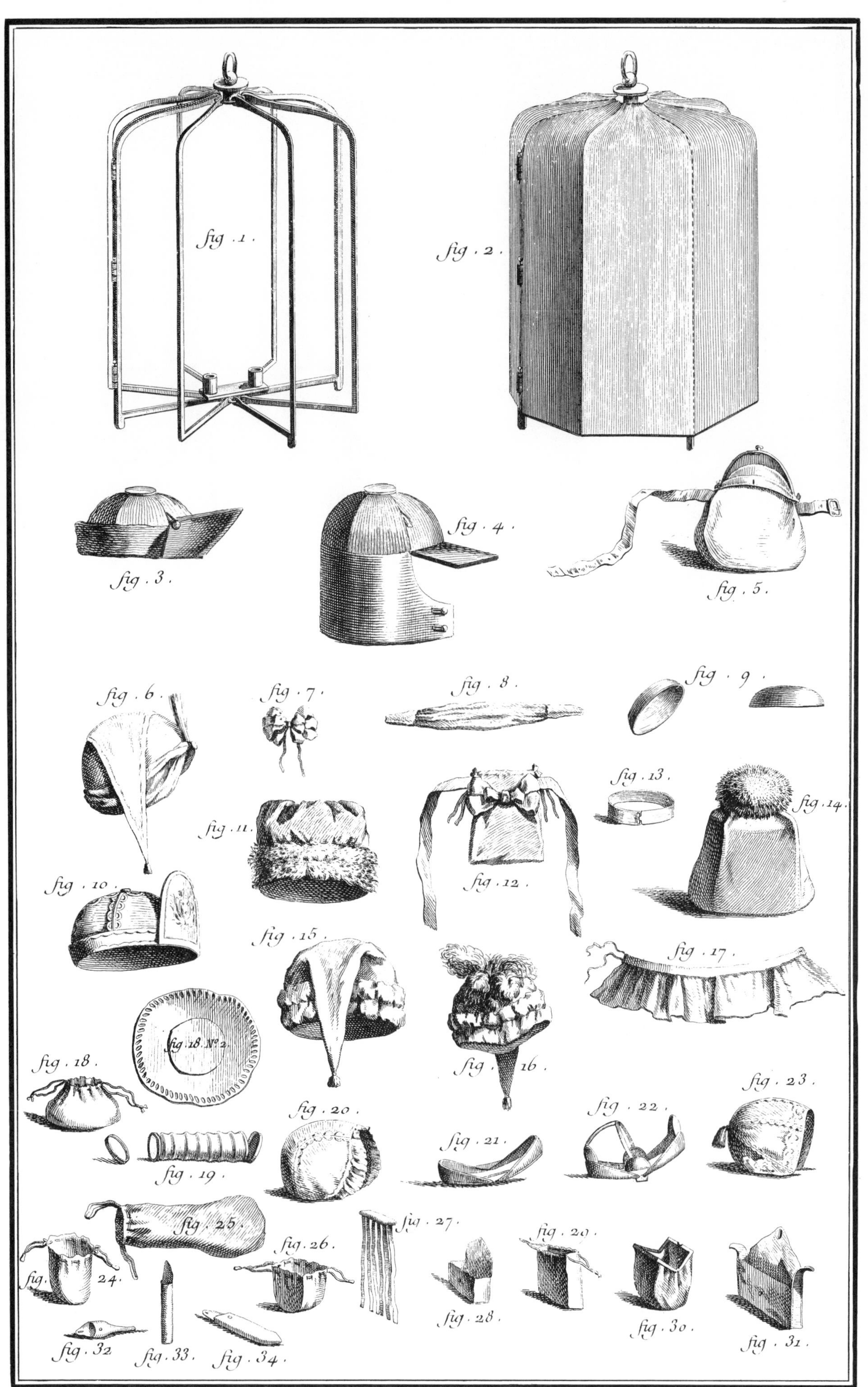
fig. 1.
fig. 2.
fig. 3.
fig. 4.
fig. 5.
fig. 6.
fig. 7.
fig. 8.
fig. 9.
fig. 10.
fig. 11.
fig. 12.
fig. 13.
fig. 14.
fig. 15.
fig. 16.
fig. 17.
fig. 18. N° 2.
fig. 18.
fig. 19.
fig. 20.
fig. 21.
fig. 22.
fig. 23.
fig. 24.
fig. 25.
fig. 26.
fig. 27.
fig. 28.
fig. 29.
fig. 30.
fig. 31.
fig. 32
fig. 33.
fig. 34.

Vol. IX, Tailleur d'habits, Pl. I.

Plate 440 The Tailor I

In the tailor shop the gentleman came into his own, for 18th century ladies did not wear suits. The tailor himself measures the client (c) while his men cut (d), heat the flatirons, and sew by the light of the window, all cross-legged on their counter.

Plate 441 The Tailor II

A man's clothes: suit and waistcoat (Figs. 1 & 2), and ordinary culottes *(Fig. 3). The* culottes *of Fig. 4 are cut in the Bavarian style. Figs. 5, 6, & 7 illustrate priests' garb: cassock, long cloak, and short cloak. Fig. 8 is a redingote, 9 a lounging robe, 10 courtdress, 11 a* gilet *(vest), and 12 a coat of the "new style."*

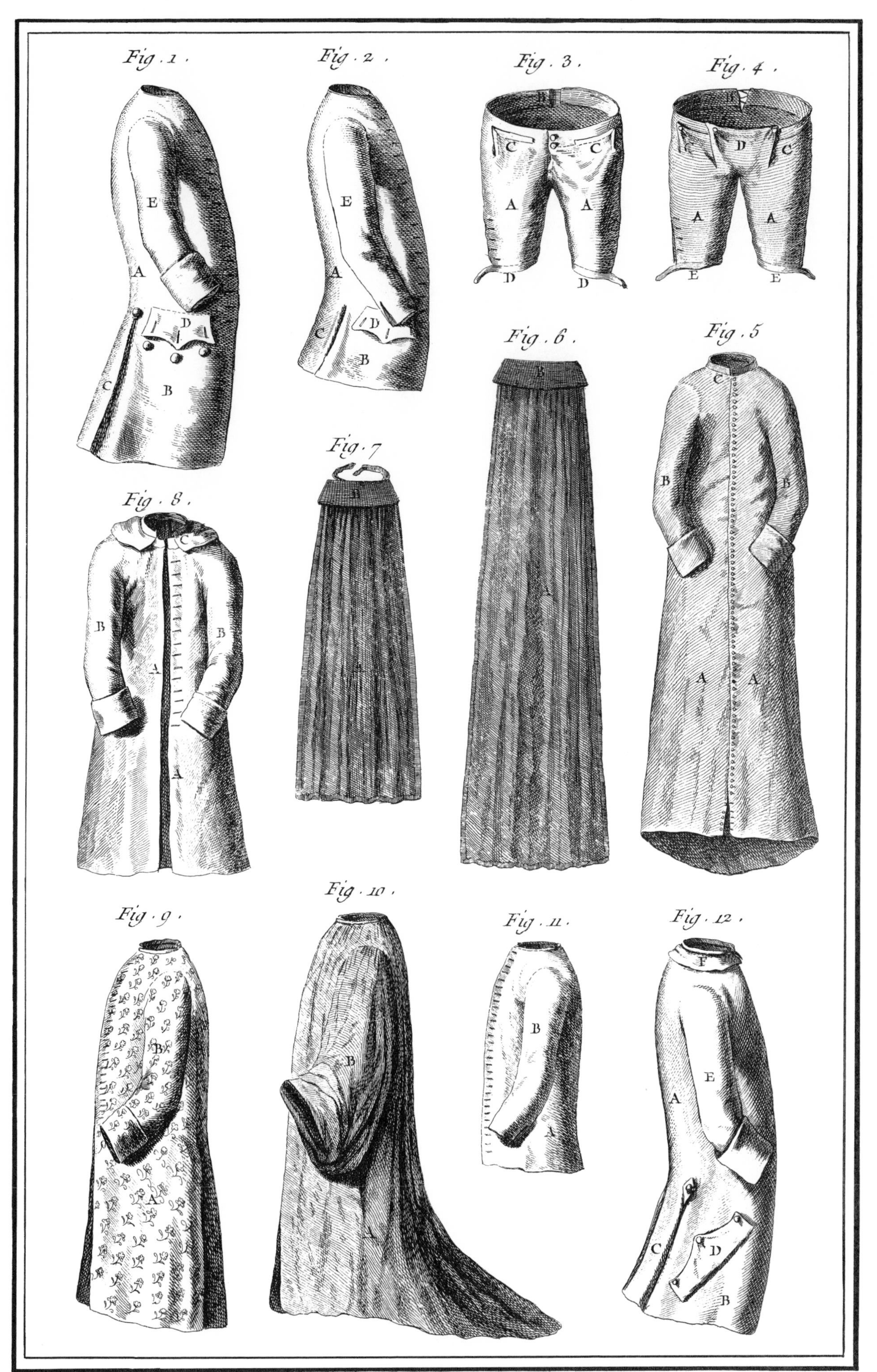
Fig. 1.
Fig. 2.
Fig. 3.
Fig. 4.
Fig. 5
Fig. 6.
Fig. 7
Fig. 8.
Fig. 9.
Fig. 10.
Fig. 11.
Fig. 12.

Plate 442 The Shoemaker

The bootmaker's is a scene of animation. Shoes and boots were made to order, of course, and therefore to fit (Fig. 1). On the wall is a rack of forms. Of the four journeymen, two seem to be gossiping (Figs. 5 & 6), one stitching a sole (Fig. 3) and one fitting a boot over a form (Fig. 4).

The shopfront is open to the roadway, and the glimpse of its cobblestone surface may, perhaps, suggest the atmosphere of an 18th century commercial quarter, the streets lined with just such shops of all trades, the doorways giving in to the court and the apartments above guarded—note how shallowly—from carriage wheels by their pylons, the dogs scavenging everywhere, the humblest tradesmen setting up their flimsy stalls in the street itself. In this case it is a cobbler (Fig. 7) who counts on the shoemaker's clients to have repair work for him.

Plate 442 The Shoemaker

Vol. III, Cordonnier et Bottier, Pl. I.

Plate 443 The Button-maker I

These people make button-molds for the casting of metal buttons. Two workers saw out a block of wood (Figs. 1, 2); a man and a woman drill into it the outline of the mold, using a bow to turn the bit where a modern carpenter would be likely to prefer a brace (Figs. 3, 4). The exact pattern is drilled out by use of the big two-man drill-wheel in the background, which by the difference in diameter of the primary and rotary wheel imparts a very rapid rotation to the drill, so rapid that a wheel of the same design could turn a lathe.

The Button-maker I

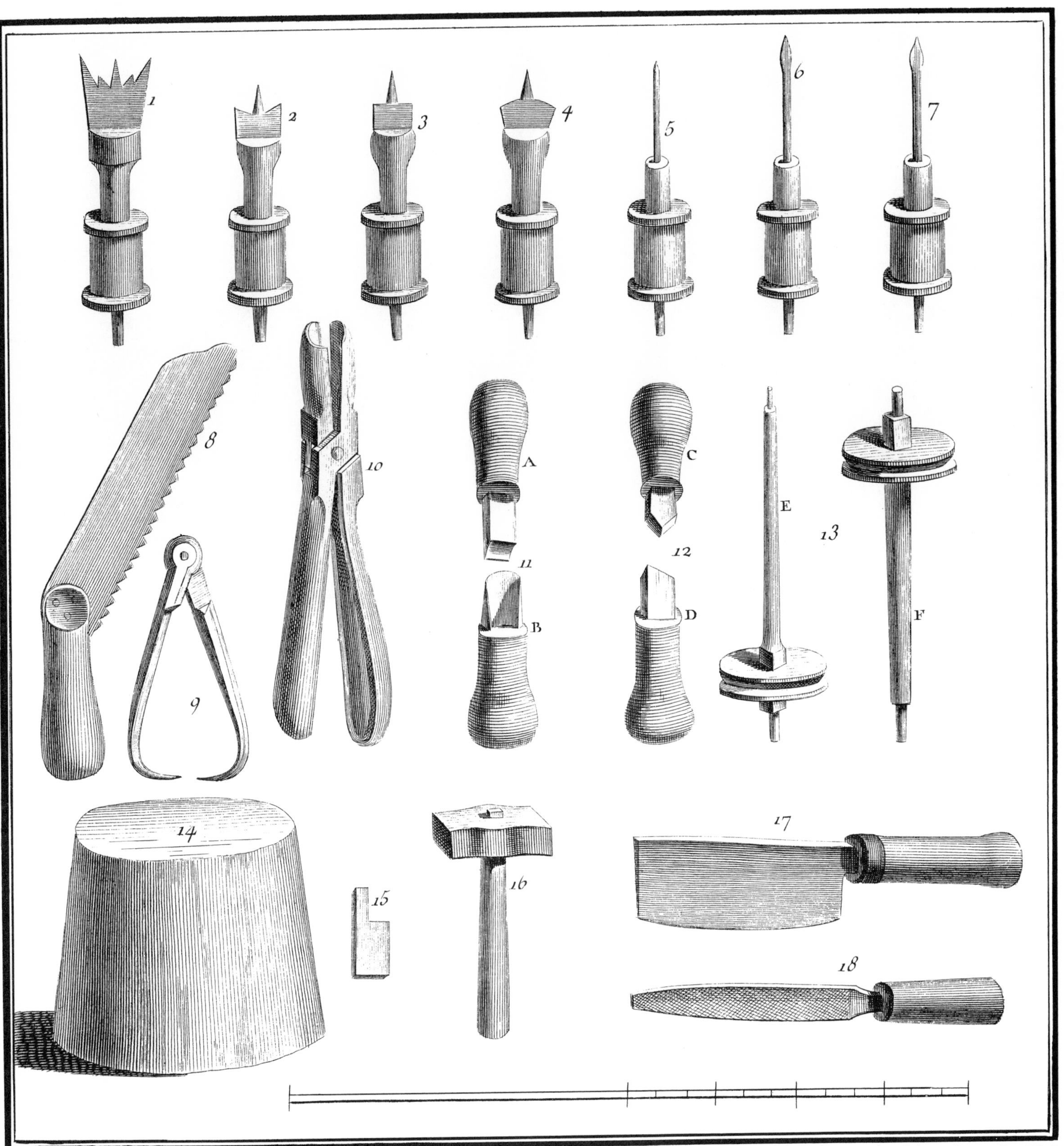

Vol. II, Boutonnier, en Métal

Plate 444 The Button-maker II

Buttons in this shop are made of resin applied to a metal shank. The artisan with his hammer raised is about to punch out a disc of metal, to be handed to the second man who immerses it in a little tub of resin kept molten on the brazier which sits before him on the table, and then presses in the design from an appropriate mold. When cool, the buttons are buffed and polished at the window (Fig. 3).

Plate 445 The Button-maker III

Vol. II, Boutonnier Passementier.

Plate 445 The Button-maker III

Here the products are covered buttons and similar fancy work. Buttons are covered with silk (Fig. 1) and stitched with piping or other raised figures (Fig. 2). At the right, a bodice is braided (Fig. 3) and a hemisphere braided ornately on a form (Fig. 4).

Vol. II, Broder

Plate 446 Embroidery

It is doubtful whether the technique of embroidery has changed for the better since the 18th century, and even more doubtful whether there are women in the world capable of doing work of this quality.

Miscellaneous Trades

Miscellaneous Trades

The modern student tends to reduce the trades of old into groupings which reflect his own ideas of the relations of production. This is an aid to organization but not, perhaps, to understanding, for it creates the impression of a more elaborately articulated commerce than actually existed, not only more highly developed in organization but on a larger scale. It is just as well, therefore, that a number of the trades illustrated in the *Encyclopedia* resist the classifications of the present volume and remain to be gathered together here under the inevitable rubric of "Miscellaneous." For it should be remembered that the vast majority of working men in 18th century Europe were peasants tilling the soil by methods more archaic than any in our first section. And, of the town-workers, only another minority would be employed in heavy industry or skilled artisanry. The larger portion, both of independent tradesmen and laborers, would get their livings in a veritable kaleidoscope of trades, from which the following are a random selection.

Vol. IV, Fleuriste artificiel, Pl. I.

Plate 447 Artificial Flowers

Artificial flowers brought out all that was most Parisian in 18th century craftsmanship, and indeed it is still possible in Paris to buy flowers which improve on transient nature by fixing the perfection of its designs in leather, glass and wax. Beauty without spot, they are better than the real thing, for they do not fade.

To begin at the left, the first woman (a), the arranger, makes bouquets. Beside her a boy (b) stamps out vellum leaves or petals. Buds and stems come from the middle of the table (c), and the machine at the far end is used for crimping and fluting materials. The man in the foreground winds iron wire with green vellum. At the far side of the big table, a girl apprentice gets instructions from both sides.

Plate 448 The Balance-maker

Vol. II, Balancier, Pl. I.

Plate 448 The Balance-maker

There was no standardization of weights and measures in the old France. (This, indeed, is one of the modern uniformities of which the most sentimental romantic could scarcely complain.) One's own scales, therefore, were a necessity. The balance-maker's appears to have been a lively shop. A lady (Fig. 4) with her scales to be repaired has at best half the attention of the workman (Fig. 2) who is filing a beam. Behind him another (Fig. 3) tests the balance of a pair of scales, and in the center a third melts on a brazier a spoonful of lead with which he will cast a weight.

Plate 449 The Bakery

Bread was the staple food of 18th century France, a heartier more nourishing bread than any to be had in modern America. It has been calculated that the average laborer or peasant in the fields required two to three pounds a day to sustain him in his work. No trade was more strictly regulated than the baker's, for none was so essential to public health and order—it was in the form of a bread riot that the women of the slums stormed out to Versailles in the first year of the Revolution to capture and bring Louis XVI, Marie Antoinette, and the Dauphin, "the baker, the baker's wife, and the baker's boy," back to Paris where the people and its representatives—or the mob and its leaders —could guarantee his concern for its wishes and its stomach.

Traditionally bakers worked naked, or nearly so. The dough is kneaded (Fig. 1) in a wooden trough, formed into loaves (Figs. 3 and 4), and baked in the oven (Fig. 5), to be sold retail. The baker was shopkeeper as well as breadmaker. The Encyclopedia *traces (somewhat dubiously) the history of the trade right back to the regulations of the Roman Empire.*

There were still three classes of bakers in 18th century Paris, the Bakers of Paris, of whom there were only 12 and whose privileges had been granted by Saint Louis; the bakers of the faubourgs, licensed by the city and subject to the same regulations; and the "foreign" bakers from outside Paris, who were licensed to sell bread at the great

Vol. II, Boulanger.

outdoor markets held on Wednesdays and Saturdays—provided that the baker or his wife was always in the stall, that he agreed to furnish a certain minimum at every market, and that he sold at a fixed price until noon. After that hour he could lower it if he needed to do so in order to sell his quota.

Plate 450 The Patisserie

Surely no more need be said than that this is a patisserie. *But notice the details—the boy beating whites of eggs (3); the sculpturing of the icing (2); the game pies that must have been one of the staples of the 18th century pastry cook. Rabbits and quail are no longer to be seen in a patisserie, but there persists in full vigor the ritual according to which some flaw detected by the customer is indignantly repudiated by the proprietor.*

Plate 450 The Patisserie

Vol. VIII, Patissier, Pl. I.

Plate 451 Starch

Starch was extracted from grits of wheaten flour too coarse to pass the baker's sieve. This process is completely obsolete, but is interesting as an illustration of the detail entered into by the commercial regulations of old France.

The starch-maker could use only wheaten remnants or rotten wheat, never good wheat. Well water (Fig. 1) is poured over the wheaten remnants in a suspension of leavening material, and the barrels allowed to stand (e, f, g) "the statutes say . . . for three weeks." After this an oily substance will have floated to the top, to be discarded. Then the suspension is filtered (Fig. 2) through a sieve (l) and rinsed three times with two buckets of water at each rinsing. The residue is itself washed carefully by stirring (Fig. 3) to be used for cattle-feed, and the pure starch solution (a, b) set aside for three days.

*After this, the starch will crystallize out, and the water will be skimmed off (Fig. 3 bis) to be used as leavened water—*eau sure*, reliable water—in future runs. Failing a supply of* eau sure*, it must be borrowed from a friend. Now the starch from the bottom of the barrel (h) is packed in baskets (Fig. 4), to be stored in the attic.*

The next two operations are to be imagined as occurring in the attic. The baskets are dumped out—each contains about 120 pounds of starch—the cakes broken into 16 pieces, and left on the floor to dry. Next, the pieces of starch must be exposed to the air on shelves built into a window frame (Fig. 5), and finally powdered and baked at a

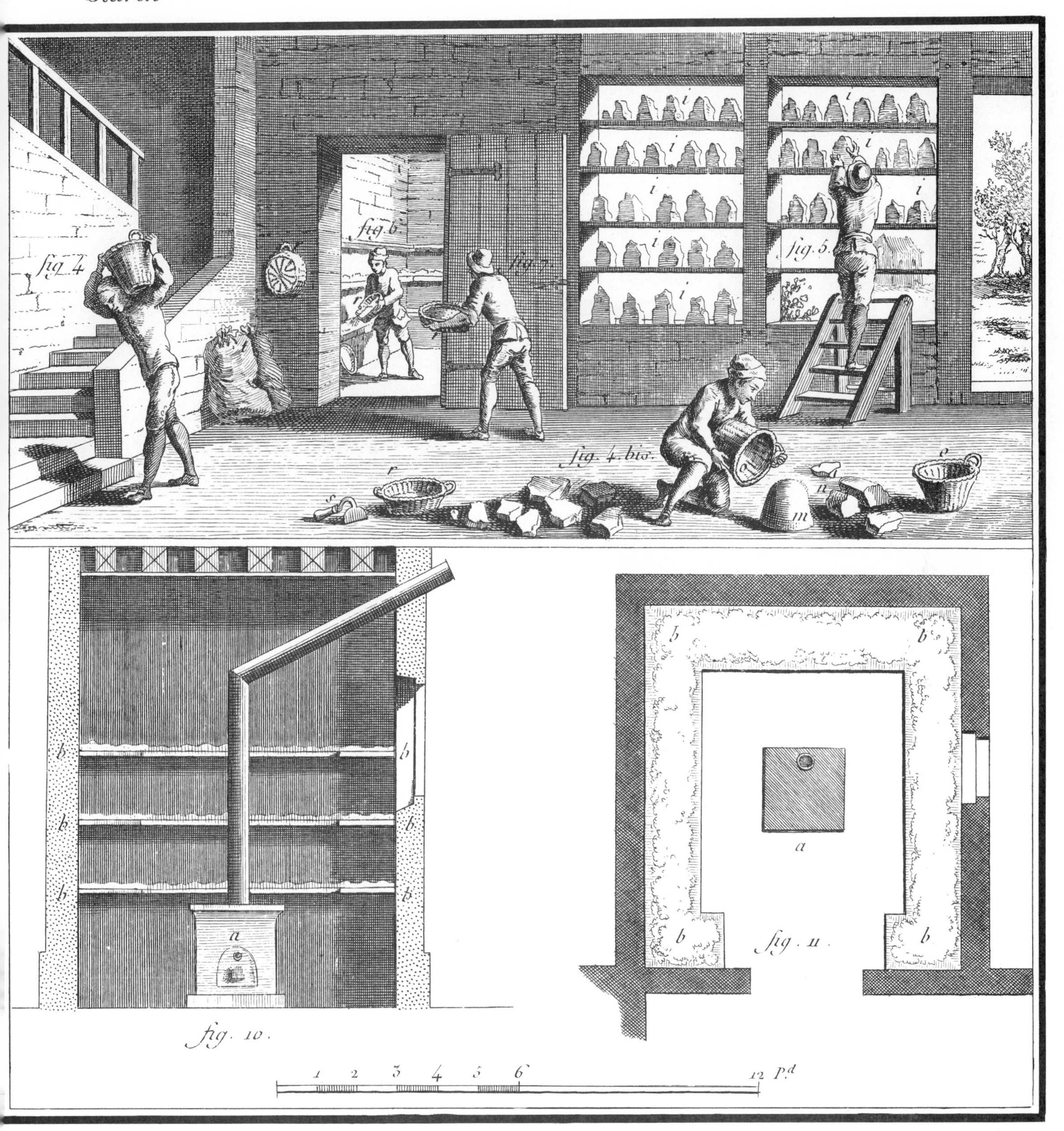

Vol. I, Amydonnier.

very low heat in a drying room (Fig. 6). "What then is starch? It is a sediment of spoiled wheat, or of wheat seconds, out of which are made a kind of white and friable dried paste."

Plate 452 Cognac

The French call brandy eau-de-vie—*water of life, and this usage gives us the right, therefore, to describe this view of a distillery as a glimpse into the heart of French civilization, the source of one of its greatest material and spiritual products. The wines for the finest brandies are grown in the district of Cognac, north of Bordeaux, where use of the old-fashioned pot-still like the one in this plate is still preferred. The tawny color and strong silky flavor of cognac is the consequence of its having been aged in oaken casks of the region.*

Cognac

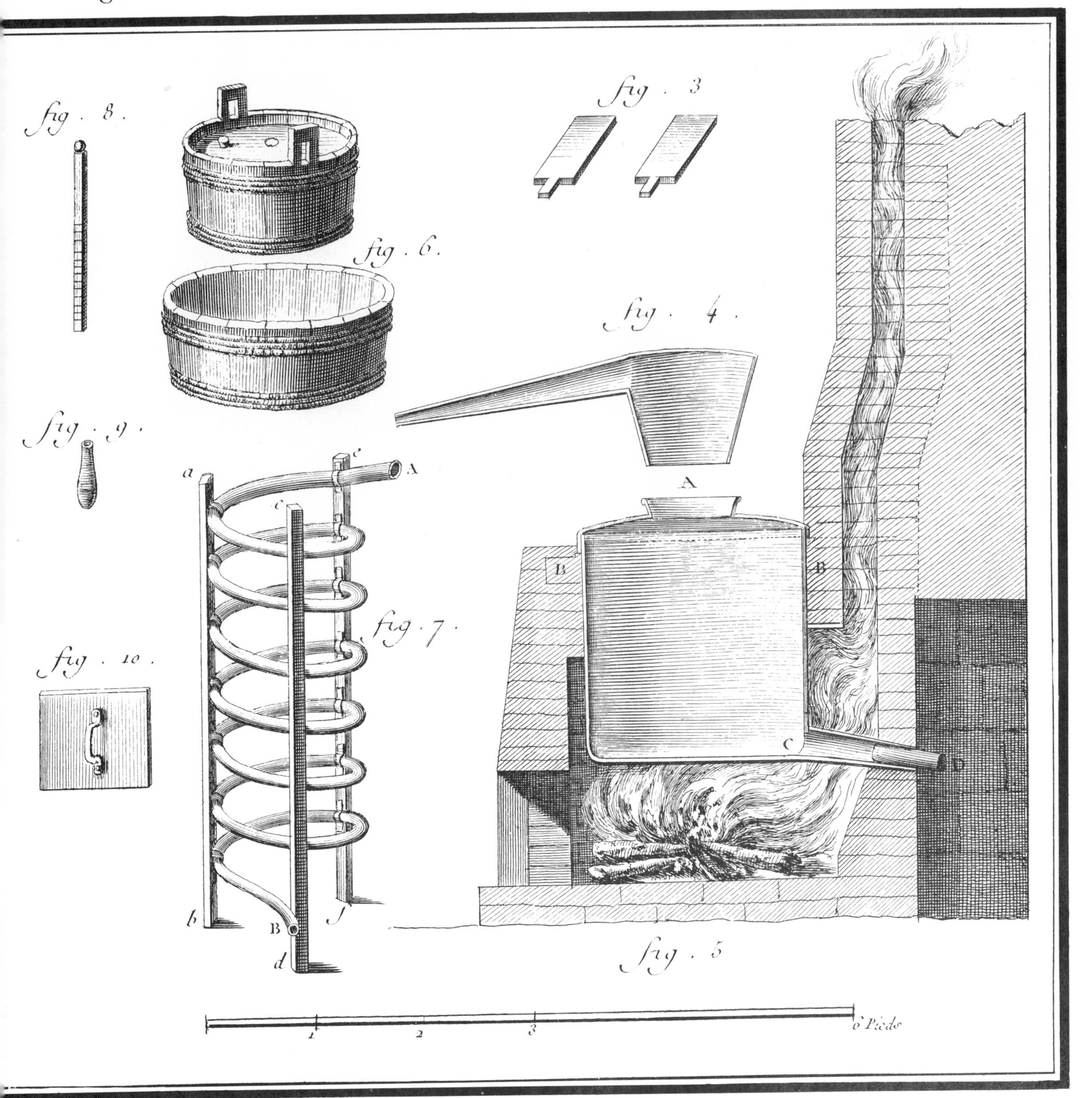

fig. 8.
fig. 3
fig. 6.
fig. 4.
fig. 9.
a
e
A
c
fig. 7.
fig. 10.
A
B
B
C
D
b
B
f
d
fig. 5
1
2
3
6 Pieds

Vol. II, Bouchonnier.

Plate 453 The Corkmaker

The corkmaker's shop illustrates the combination of petty manufacturing and merchandizing characteristic of 18th century commerce. The owner (Fig. 1) and his apprentice (Fig. 2) cut cork stoppers, while the owner's wife sorts out sizes and sells to passing customers. The shopfront is open to the street, and of course to the weather, and Madame, garbed in decent bourgeois modesty, wears a shawl against the draft.

Vol. II, Chandelier, Pl. I.

Plate 454 Candles

Readers who have visited colonial Williamsburg will recognize the candlemaker's shop: cutting the wick (Fig. 1), melting the tallow (Fig. 2), dipping the wick (Fig. 3), and pouring the candles in molds (Fig. 4). It is easy to see how candles were made. What is difficult to remember is how little light they gave, and how dim was life by night. It was the people in the 18th century salons who shone in brilliance, not the lighting.

Vol. IX, Tabletier-Cornetier, Pl. I.

Plate 455 Horn

In the 18th century, things like combs, brushes, checkers, chessmen, which one thinks of dimly as made out of horn or bone, actually were made of these materials. Chemistry could not then imitate them. Working in horn is the trade of this shop. Its instruments are heat, either on a grill (a) or at the fireplace (b), which softens horns; mallets, cleavers and chopping block (g, h); and most important of all a long multi-action press (f) used to mold the softened horns into desired forms.

Vol. IV, Gainier, Pl. I.

Plate 456 The Casemaker

Scabbards were among the chief products of the casemaker, who specialized in luxury containers of one kind or another. The jig saw (Fig. 1) is his most important tool, and his men had to develop the sort of neat-fingered skill required (for example) to line cases with fabric (Figs. 2, 3, & 5), while the master ranges a finished violin case on the shelf.

Vol. II, Boisselier, Pl. I.

Plate 457 The White Cooper

This is a busheler's or white cooper's shop. It produced all sorts of small containers: drums, lanterns, bushel measures, pails, wooden shoes, small measures, bellows, sieves, and colanders. The workman in Fig. 1 is shaving a bottom for a bushel measure. The measure will be bound together with hoops like those on the wall beneath the drum in Fig. 3. The other workman makes a bellows airtight.

Plate 458 The Cooper

Vol. X, Tonnelier, Pl. I.

Plate 458 The Cooper

Many small craftsmen did not require elaborate equipment, and their margin was so slender that they would not have found it possible to support the overhead of renting a shop. In such a case they might for a small fee acquire the right to practise their craft under the arches of the gallery of some public building or large hôtel particulier. *Or perhaps they might inherit a privilege, sometimes over many generations. Such is the arrangement of this barrel-maker, who disposes of each day's product as he makes it and whose men take their tools home with them. All he has to do is secure his benches back against the wall when night comes.*

This is no fashionable quarter. Windows are left unglazed. Sentimentalists who deplore the destruction wrought by Haussmann in the 19th century, when the great boulevards were cut through quarters like this, might imagine what Paris would be if it had to sift the life of a modern city through such a sieve of evil alleys.

Vol. X, Vannier, Pl. I

Plate 459 The Basket-maker

The basket-maker is worse off than the barrel-maker. He works and probably lives in this cellar, where even more miserable employees adapt basketwork to a far greater variety of uses than it has today. They even make a figure of it—sculpture which was the artistic equivalent of the cigarstore Indian in America.

Plate 460 Pottery

Vol. VIII, Potier de terre, Pl. I.

Plate 460 Pottery

The potter's shop is rather a miscellany. The man in the right foreground (b) is working on a base for a chafing dish, which is what the knobby objects are at his feet on the other side of the barrel. Clay pipes are the specialty of the workman at the table (d).

The shop turns out ceramics of all sorts. As was the case with lead, the expanding interest in chemistry produced a demand for much chemical ware. The potter himself works at his kickwheel in a hideously uncomfortable posture (a). His kiln is built right into the wall of the shop (e).

Supplement, Art de faire la Porcelaine, Pl. I.

Plate 461 Porcelain I

The better grades of porcelain were products of the Limousin, the region around Limoges, where there are peculiarly rich deposits of kaolin. The clay is very hard and had to be broken up by sledges (A), ground in a mortar (B) and sifted free of pebbles (C). Before going to the potter, the raw clay was roasted on an open grill (D) then thinned with water (G) and worked up in the shop (E) in the rear. Unfired vessels are hung on pegs (H) to await their turn in the kiln (F).

Supplement, Art de faire la Porcelaine, Pl. III.

Plate 462 Porcelain II

The main gallery here is for decorating porcelains. Paints are prepared at the table (C), applied by the artists working at a work bench placed in a good light (D), and baked on in the little oven at the rear (A).

The open door gives a glimpse into the shop where the sculptors produced shepherdesses, mythological figures, stags, unicorns, figurines, lamps, vases, all the whatnots so necessary to an 18th century interior.

Plate 463 Playing Cards I

The manufacture of playing cards was one of the most closely supervised industries of the old regime, for like tobacco and other commodities which ministered to pleasure instead of social utility, playing cards were easily and hence heavily taxed. Moreover, the administration of the tax was farmed out to a corporation of financiers whose incentives and habits reinforced each other in efficient collection.

Playing cards were made out of sheets of good heavy paper, stiffened with paste and laminated in a press (Fig. 7). After drying, the sheets—each of a size to make 20 cards—are polished (Fig. 3), printed, and colored. The basic design is printed in black ink from a single block, a step which is not shown for the reason that regulations required the printing to be carried out in the office of the tax farm, to obviate evasion of the duty of a penny a card. Fig. 5 opposite shows a single sheet of face cards in black outline. The cards are colored by hand. A different stencil was required for each color. In Fig. 1 the artisan is applying yellow to a block of face cards, using a stencil like that in Fig. 6 opposite: and in Fig. 2 his companion is painting a block of clubs, coins, cups, or swords. In the modern world, the latter three suits have become diamonds, hearts, and spades.

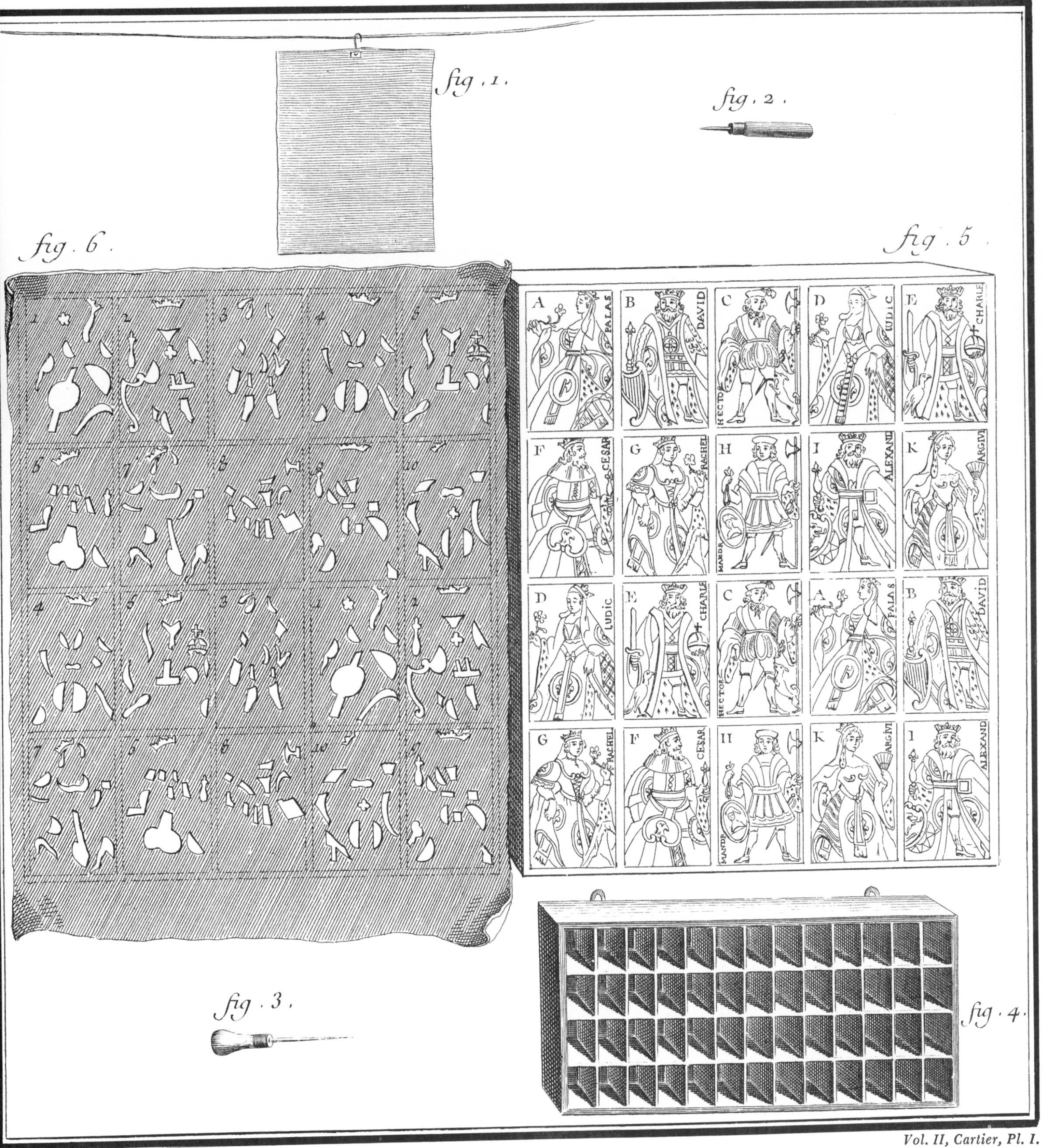
fig. 1.
fig. 2.
fig. 6.
fig. 5.
A PALAS
B DAVID
C HECTOR
D LUDIC
E CHARLE
F CESAR
G RACHEL
H MANDR
I ALEXAND
K ARGIVE
fig. 3.
fig. 4.

Plate 464 Playing Cards II

This is a close-up of the press and stove for making stiffening paste, a mixture of flour and starch.

Playing cards have been known in Europe since the early fifteenth century. Beyond that their history is a matter of legend and myth. There is, for example, the story that the four suits represent the four seasons, the 13 cards in a suit the lunar months in the year, the 52 cards in a deck the weeks of the year, and the sum of the face values of the card, plus one for the joker, the 365 days of the year. Even more striking is the circumstance that the number of letters it takes to write the series, Ace, Two, Three and so on through Jack, Queen, King, amounts to 52. This is true not only of the English language, but also of French and German. But, unfortunately for symbolic significance, early packs contained 56 cards, a number which cannot be made to play these numerological games.

Vol. II, Cartier, Pl. II.

Vol. III, Charron, Pl. I.

Plate 465 The Wheelwright I

Among trades that have virtually disappeared is the wheelwright's. Trimming segments of the rim out of solid blocks with axe (Fig. 1) and adze (Fig. 2) was the most laborious part of his work.

Plate 466 The Wheelwright II

Vol. III, Charron, Pl. II.

Plate 466 The Wheelwright II

All the parts of the wheel are made, and then assembled. Rim sections (Fig. 1) are mortised to receive the spokes; the spokes are mounted on the hub (Figs. 2, 3); and the sections of the rim are laid on to fit (Fig. 4). Parts are made too long for the very sound reason that it is easier to shorten than to lengthen a piece of wood. The rim is eventually mounted piece by piece on the spokes, and bound (Fig. 5) to make a solid wheel.

Vol. II, Bourlier, Pl. I.

Plates 467, 468 Harness-making I & II

These workers, who are cutting (Fig. 1), punching (Fig. 2), and stitching (Fig. 3) straps of leather, are employees in a harness-maker's establishment. Judging from the finished product hanging on the wall, he specializes in harnesses for carriage horses, which are shown in plate 468.

The rig at the top of plate 468 would be used on a pair of horses, or for the front pair of a team, that in the center for the middle pair in a team of six, and that at the bottom for the pair nearest the carriage.

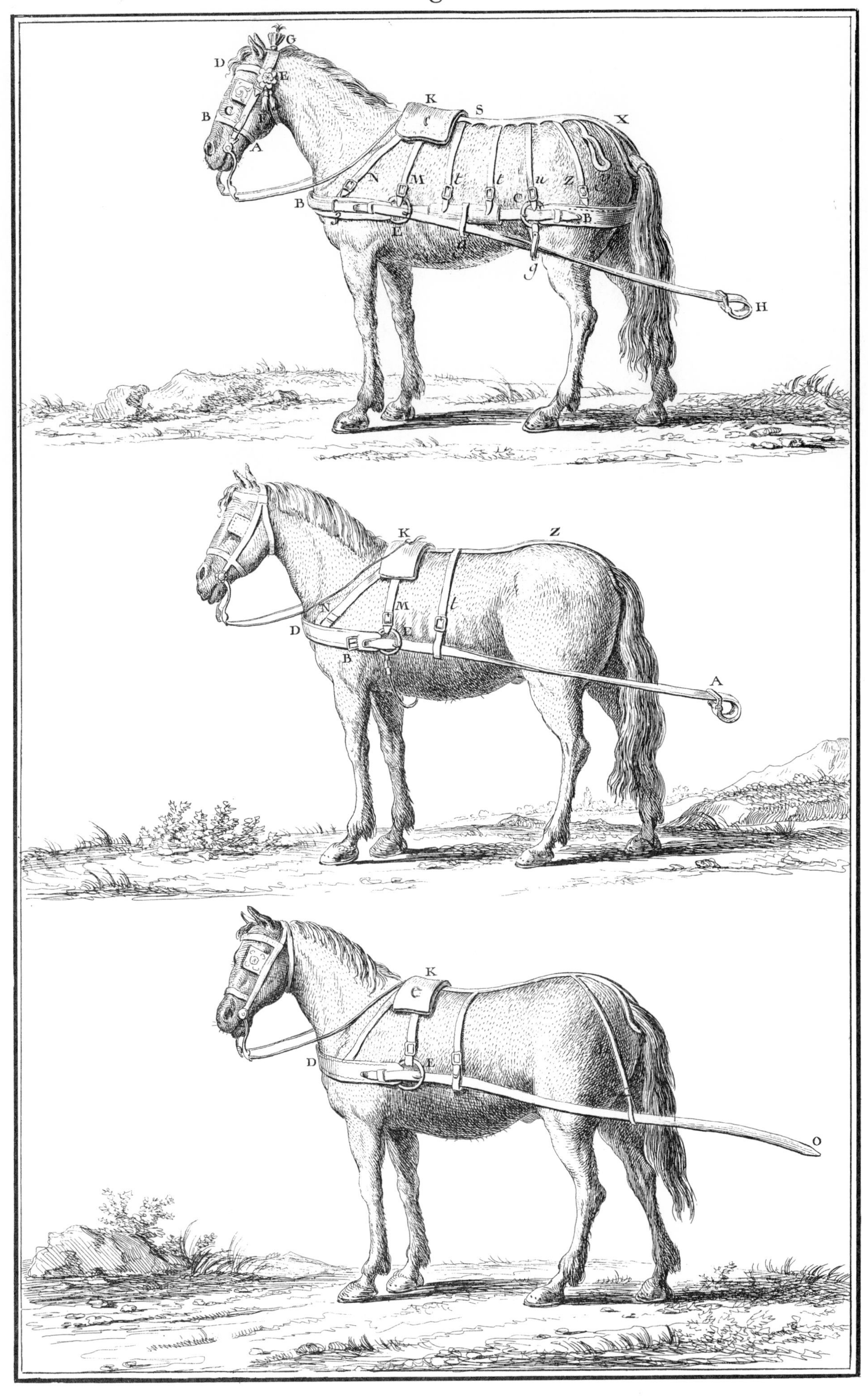
G
D
E
B
C
A
K
S
X
N
M
B
E
B
H
K
Z
N
M
D
E
B
A
K
D
E
O

Vol. II, Bourrelier, Pl.

Plates 469, 470 Saddlery I & II

Another shop makes saddles, pack saddles and horse collars, the latter two items for dray horses. It would appear from these designs that drayers rather than gentlemen formed the clientele of the establishment. The next plate illustrates various types of collars and pack saddles, which differ according to the nature of the load to be handled and show as much variety of design, perhaps, as do modern devices for pulling and hauling. To get the most out of your horse, you had to harness him properly.

A
E
D
C
B
Q
A
P
O
M
C
B
D
E
F
D
L
G
H
G
G
A
D
M
P
O
H
D
C
E
L
G
B
H
H
D
C
E
X
R
X
G
P
F
P
O
q
N
p
M
S
a
a

Plate 471 A Brewery

Horses are being used as a prime mover in a brewery, where they provide power both for the grinding mill and the hoist for grain (see the loft, above). This is deadly work, monotonous even for a horse—and it illustrates the necessity of adapting the collar to the work so as to give the horse the greatest possible mechanical advantage.

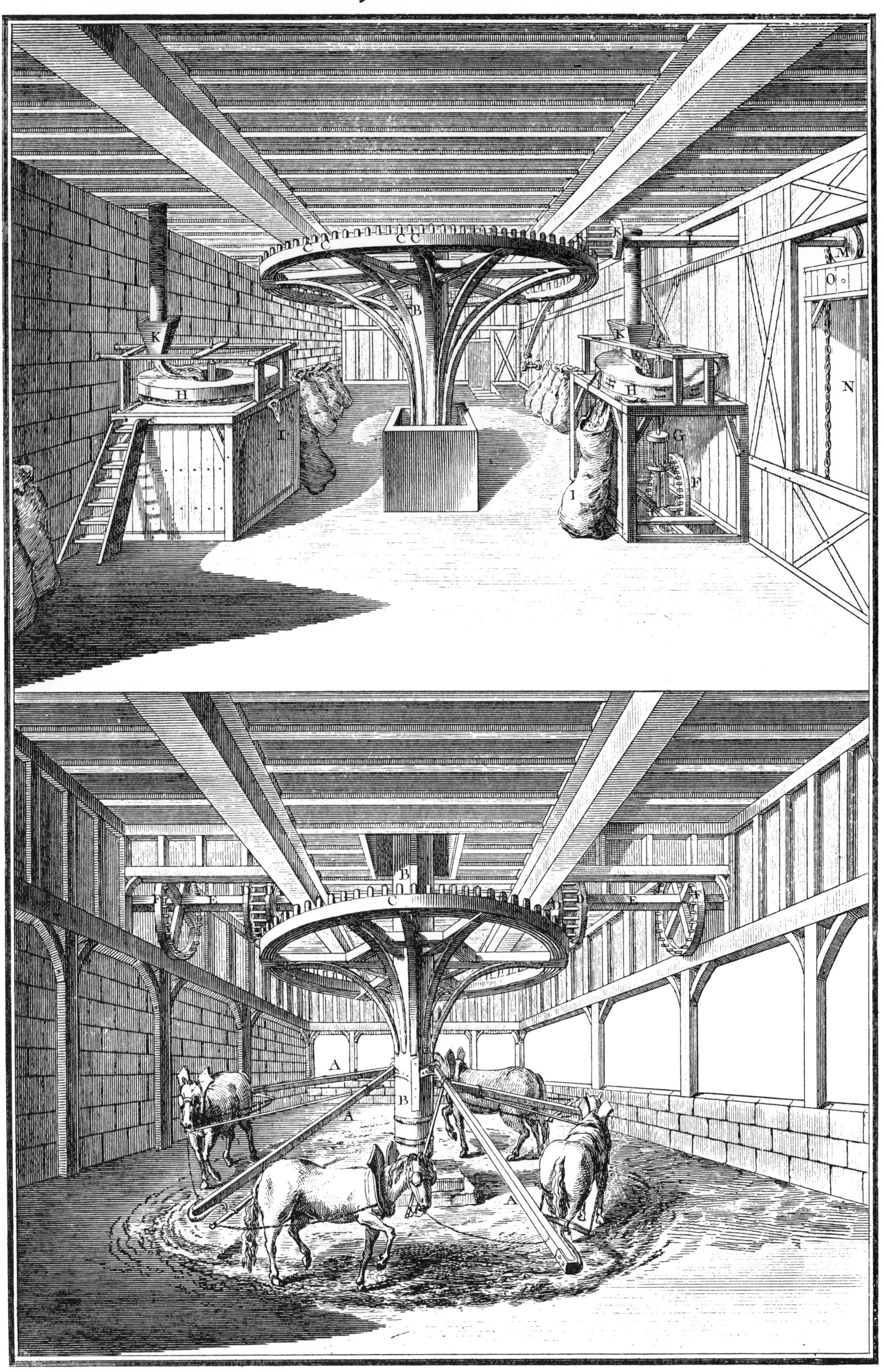

Vol. II, Brasserie, Pl. III.

Plate 472 Soapmaking I

Marseille was the center of the French soap industry because its position as Mediterranean entrepôt *gave ready access to supplies of olive oil. In northern Europe soap was made of tallow. In Latin countries olive oil was the source of the fats, which were boiled with lye to make—in the best qualities—a pure white, soluble soap smelling slightly of violets.*

We are shown the second story of the soap factory. The ground floor is taken up with furnaces, one to a soap-boiler (d) of which we see only the surfaces. The cauldrons are in fact about eight feet deep. Along the side wall are ranged tanks of lye (e), in three different degrees of strength.

The method of control in the manufacture of the lye may interest industrial chemists. Two hundred pounds of soda of Alicante (the Spanish source of the best "mineral soda," made by burning the barilla plant which flourished there) were mixed with an equal volume of freshly slaked lime and leached in a filtering vat. The resulting lye was tested with an egg. As long as the egg floated end up, the lye was "strong." When it turned on its side, the concentration was "medium"; when it barely floated, like an iceberg, the lye was weak; and when the egg sank, the remaining lye was worthless.

Vol. IX, Savonerie, Pl. I.

Plate 473 Soapmaking II

Soap-boiling was a slow process. Lye was added to the oil in a gingerly fashion (top, Fig. 2), starting with the weakest grade. As saponification progressed it might be necessary to enrich the mixture with further oil (Fig. 4). After two or three buckets of strong lye, the mixture began to look homogeneous. Then it was cooled and the surface allowed to harden.

Vol. IX, Savonerie, Pl. II.

Plate 474 Soapmaking III

The soapy matter was skimmed off and reboiled after emptying the boiler. After about two hours, it reached the consistency and took on the appearance of honey. At this point the fires were stopped again and the soap allowed to cool for a day. Still barely liquid it was dipped out of the boilers (top, Fig. 3), and transferred to rectangular tanks where it hardened in great cakes (Pl. 474). These were sliced horizontally by drawing an iron wire through them (Fig. 3) and then vertically (Fig. 2) by a sort of soap cleaver.

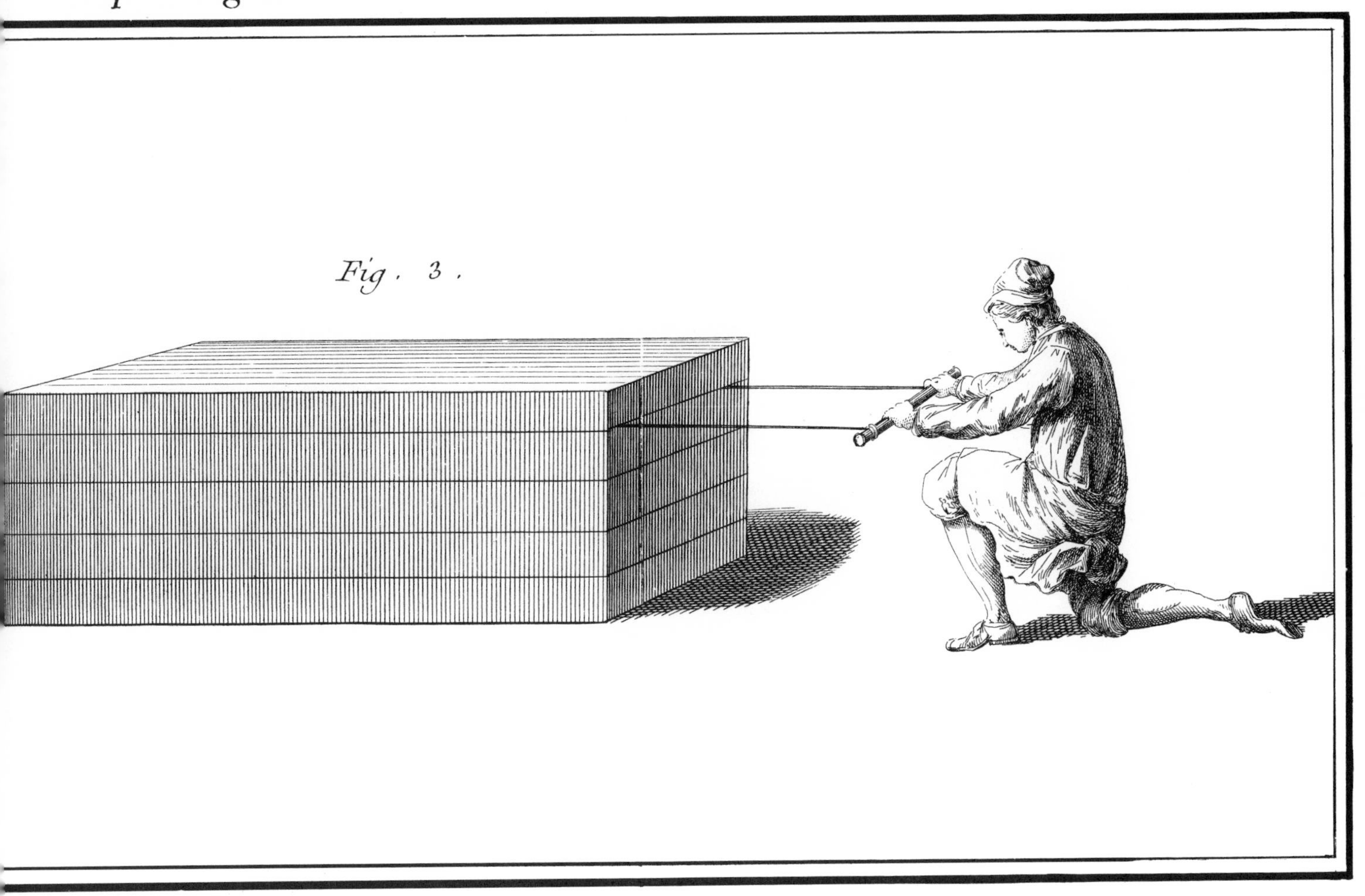

Vol. IX, Savonerie, Pl. IV.

Plate 475 Ropemaking I

Ropemakers needed room for their craft, for the fewer splices in the rope, the stronger it was. Some ropewalks were over 800 feet long. This plate shows a much smaller shop, a corderie du roi, *where cordage was manufactured to naval specifications.*

The process begins with hanks of carded, well-combed, clean hemp fiber (l, at right). The spinner gathers up a bundle of fiber sufficient to spin a strand the length of the ropewalk, and fastens a loop of it to a wheel hook (B, at the right) on the spinning wheel. He supports the hemp behind his back with one arm, and pays it out with the other as he backs away from the wheel.

As the spinner moves away, the operator of the wheel turns a crank which causes all the little wheel-hooks on the top of the wheel to revolve rapidly. This twists the fibers into a yarn, the uniformity of which depends upon the spinner's keeping an even pace. As the yarn grows longer, the spinner leads it over supports (G) or hooks suspended from the ceiling (G). When he has used all his hemp (extreme right) he calls back to the wheel man, who detaches the yarn from the wheel and splices it to the end of the continuous roll (i) which is being reeled up for storage.

This plate shows only a few yarns in progress. In actual practice, however, the foreman begins with the top hook on the wheel. When he is four arm-lengths away, another spinner starts the next hook, and so on until all the hooks are filled. A good crew could turn out, in a day, 700 pounds of spun hemp.

A
D
C
K
fig. 2.
fig. 4.
fig. 3.
fig. 7.
fig. 5.
fig. 6.
fig. 1.

G
H
G
l
B
E
fig. 12.
u
t
s
x
fig. 13.
u
y
t
y
fig. 9.
fig. 14.
fig. 15.
7
8
6
fig. 16.
9
5
4
fig. 8.
o
o
o
p
fig. 11.
q
p
r
o
fig. 10.
n
n
12 Pieds

Vol. III, Corderie, Pl. I.

Plate 476 Ropemaking II

Further operations in ropemaking arise from a characteristic of twisted fibers: they will unravel unless controlled by a twisting in the opposite direction. The yarn manufactured in the previous plate, for example, would unroll if cut off in short lengths. To counteract this tendency yarns were twisted together in the opposite direction to the individual twists in each yarn. That is, two or more yarns wound clockwise would be twisted together counterclockwise to form a strand. This principle of balanced twistings is carried on up through ropemaking to produce rope that is "neutral".

In this illustration yarn is being twisted into twine of two strands (Fig. 3) and merlin *of three (Fig. 4). In either case one end of a hemp yarn is attached to a hook on a spinning wheel (Figs. 1, 2), while the other end is fastened to a hook in the wall or to a weight (c, in the center of the windows). As many yarns are set up as are desired, and then they are "laid," or twisted against the previous twist. The wheel operates on the same principle as the larger spinning wheel of the ropewalk, but it is aided by a "top," or zipper-like collar (Figs. n, 4), which keeps the strands even and spreads the torsion uniformly.*

Plate 477 Ropemaking III

Here strands are being twisted into rope, and ropes into cables. The middle vignette shows the preparation of a cable-laid rope formed out of three ropes. The bottom illustration shows a four-rope cable which is shroud-laid, or wrapped around a central strand (Fig. 7 bottom). The geometric figures at the bottom show cross sections of different kinds of cables. The variety of ropes required on shipboard was legion, and the ways of combining ropes and strands to give different characteristics were limited only by the geometrical possibilities.

Plate 478 Ropemaking IV

After laying, cordage destined for naval use had to be tarred to prevent its rotting. This was a far more cumbersome job than might be supposed, and required a rather elaborate construction.

The fire (C) melts tar in a square cauldron (A) through which cables are paid, and it also heats the great ovens (G, F) in which the cables are coiled to dry. This heating serves not only to dry the rope but also to impregnate all the fibers with protective tar. When the men are finished, they will have filled the whole space with a coil of cable which will cook for a day or two.

The cross section at the lower right shows the arrangement of furnace, cauldron, and ovens. This structure contained four such ovens, of which only the two front ones are shown.

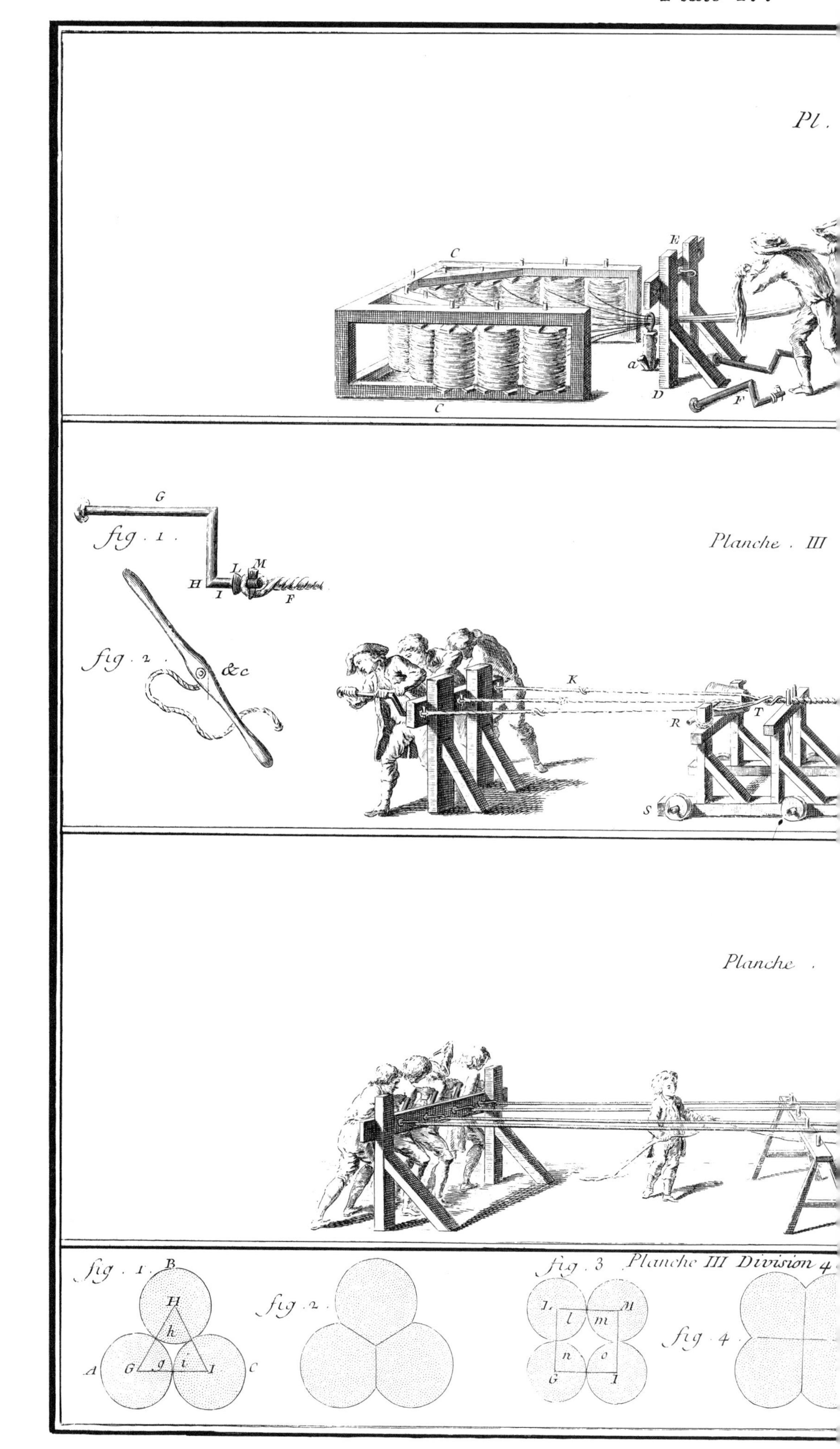
Pl.
C
E
a
D
F
C
G
fig. 1.
H
I
L
M
F
fig. 2.
&c
Planche . III
K
T
R
S
Planche .
fig. 1. B
H
h
A
G
g
i
I
C
fig. 2.
fig. 3 Planche III Division 4
L
M
l
m
n
o
G
I
fig. 4.

1.ere
O
N
b
a
Q
c
fig. 5.
P
R
T
R
T
fig. 3.
fig. 4.
X
Q
Y
V
ion 3e.
fig. 6.
fig 7.
B
a
a
A
a
a
fig. 8.

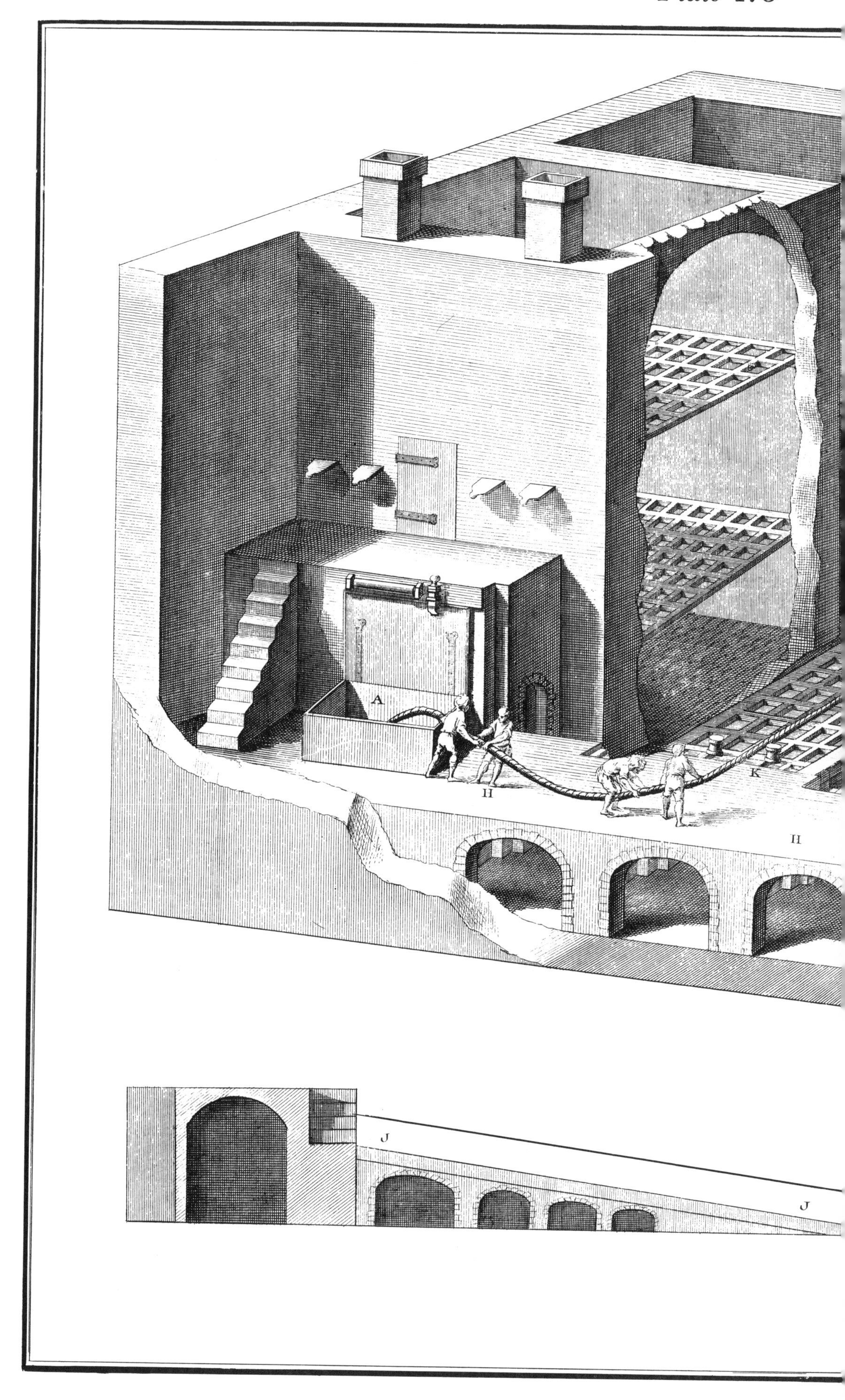
A
H
K
H
J
J

G
F
A
C
D
G
H
D
F
A
D
B
E
C

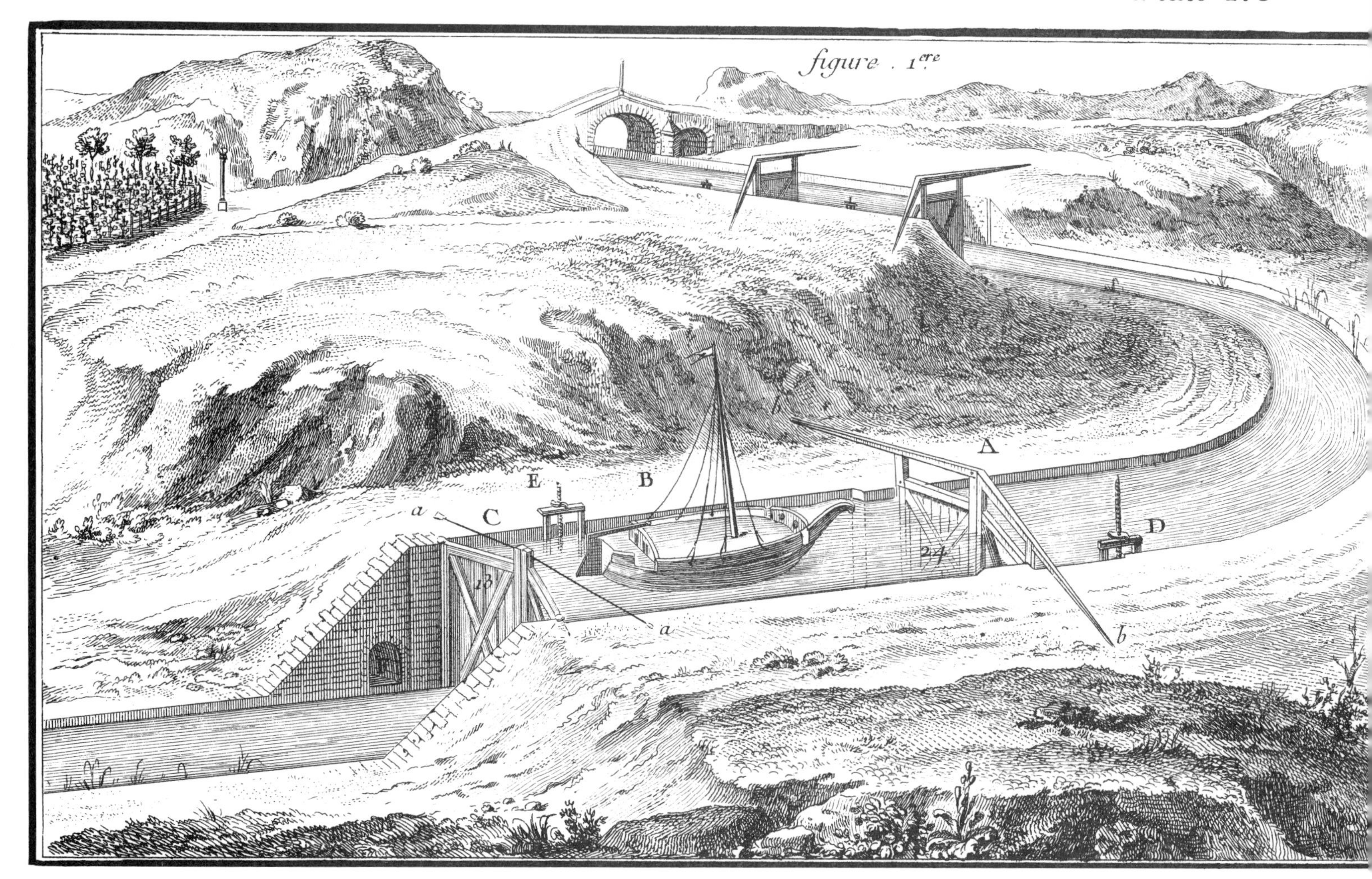

Plate 479 Canals

The French built the first great canals of the modern world. The Briare canal linking the river systems of the Seine and the Loire was opened in 1642. Beset by delays and difficulties, its construction took fourteen years. From its southern terminus at Briare it carried vessels up 128 feet a distance of eight miles to the divide between the valleys of the Trezée and the Loing and down 266 feet over a distance of 23 miles to Montargis, where it joined the watershed of the Seine.

The experience gained in building the Briare was put to work in the construction of the Languedoc Canal, opened in 1681. This struck the imagination of the world. It joined the Mediterranean to the Atlantic across a summit which is 620 feet above the Mediterranean entrance at L'Etang de Thau and 206 feet above the Garonne at Toulouse. The channel was fifty feet wide and the canal 150 miles long. In celebrating the works of Louis XIV, Voltaire placed it before the Louvre and Versailles: "The most glorious of his monuments by its utility, its magnitude, and its difficulties, was this canal of Languedoc which joins the two seas."

The canal of Languedoc contained 100 locks. The plate illustrates the engineering problems involved in constructing such a "summit" canal, which leads water across

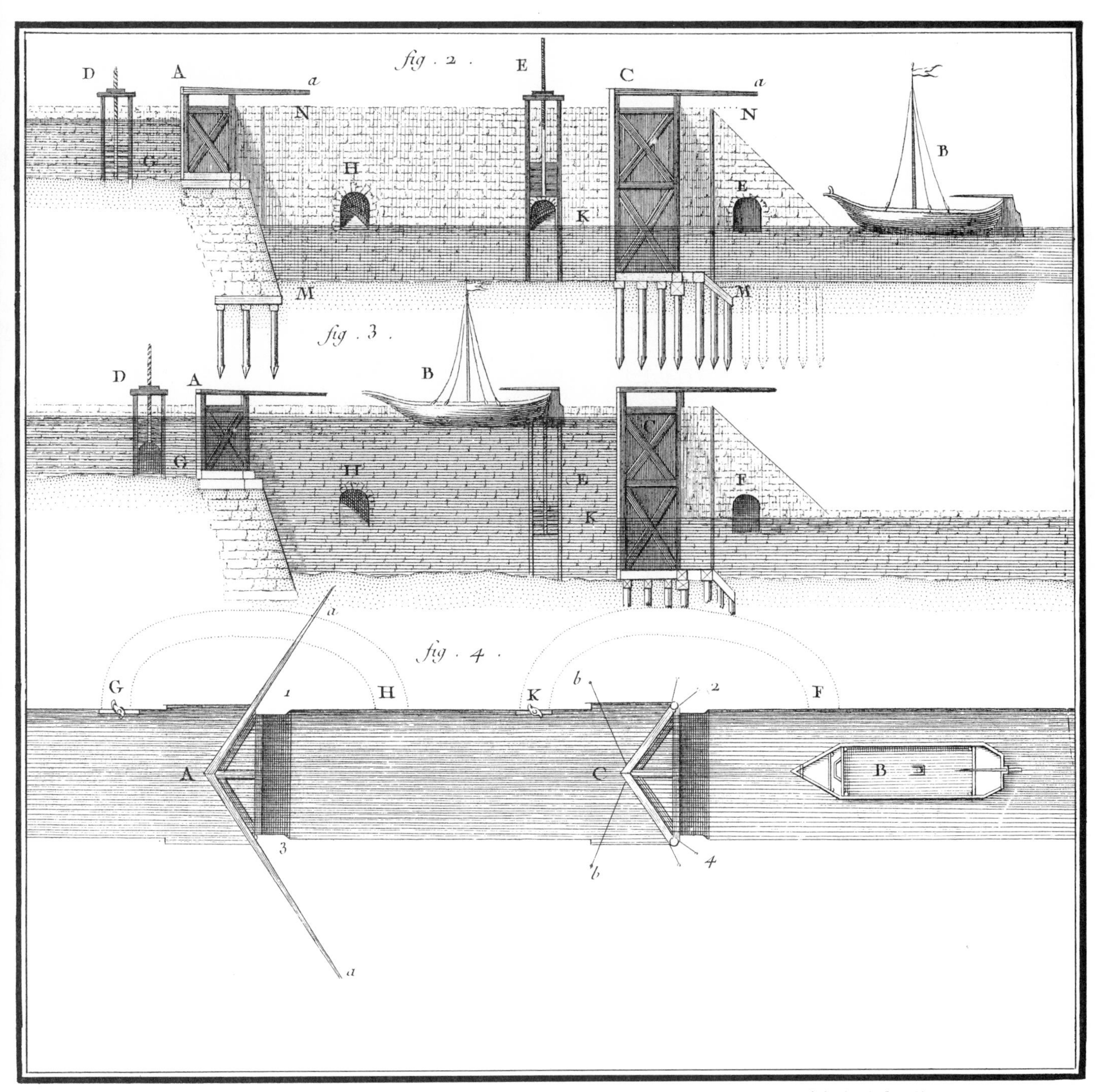

Vol. V. Hydraulique, Canal et Ecluses.

the rising ground instead of through a tunnel. A and C represent lock gates. D and E on the top figure are sluice gates for raising or lowering the water level within the lock. F is the sluice for draining. The vignette above, Fig. 4, shows in dotted lines the conduits from one chamber to the next.

Canals multiplied in the 18th century. Unable to imagine the railroad, economists looked to the development of water transport to carry the steadily increasing load of commerce and manufacturing. The river and canal system of western Europe still carries a far larger proportion of commercial traffic than in America.

Plate 480 The Confectioner's I

In the ancient world and in Renaissance Europe candy had been made mostly of honey. Confectionery of the modern type developed with the sugar industry (see Plate 37ff)."It seems as if this art had been invented only to flatter the taste, and in ways as various as its products. There is no sort of fruit, no flower and no plant, no matter how delicious it may be in nature, to which candying cannot lend a flavor more seductive and agreeable. It sweetens the bitterness of the sharpest fruits and turns them into sweetmeats. Confectionery provides the tables of great lords with the loveliest of ornaments. For it can execute all sorts of designs and figures in sugar, and even achieve considerable works of architecture."

The workroom is described in the Encyclopedia *as a laboratory. In Fig. 1 a workman puts little candied fruits to drain on the grill of a mold, and another (Fig. 2) browns almonds in sugar to make pralines. The stove is really a small scale sugar boiler. On the wall behind it hangs a strainer of red copper, and in the left rear syrup is dripping through a cheesecloth sieve.*

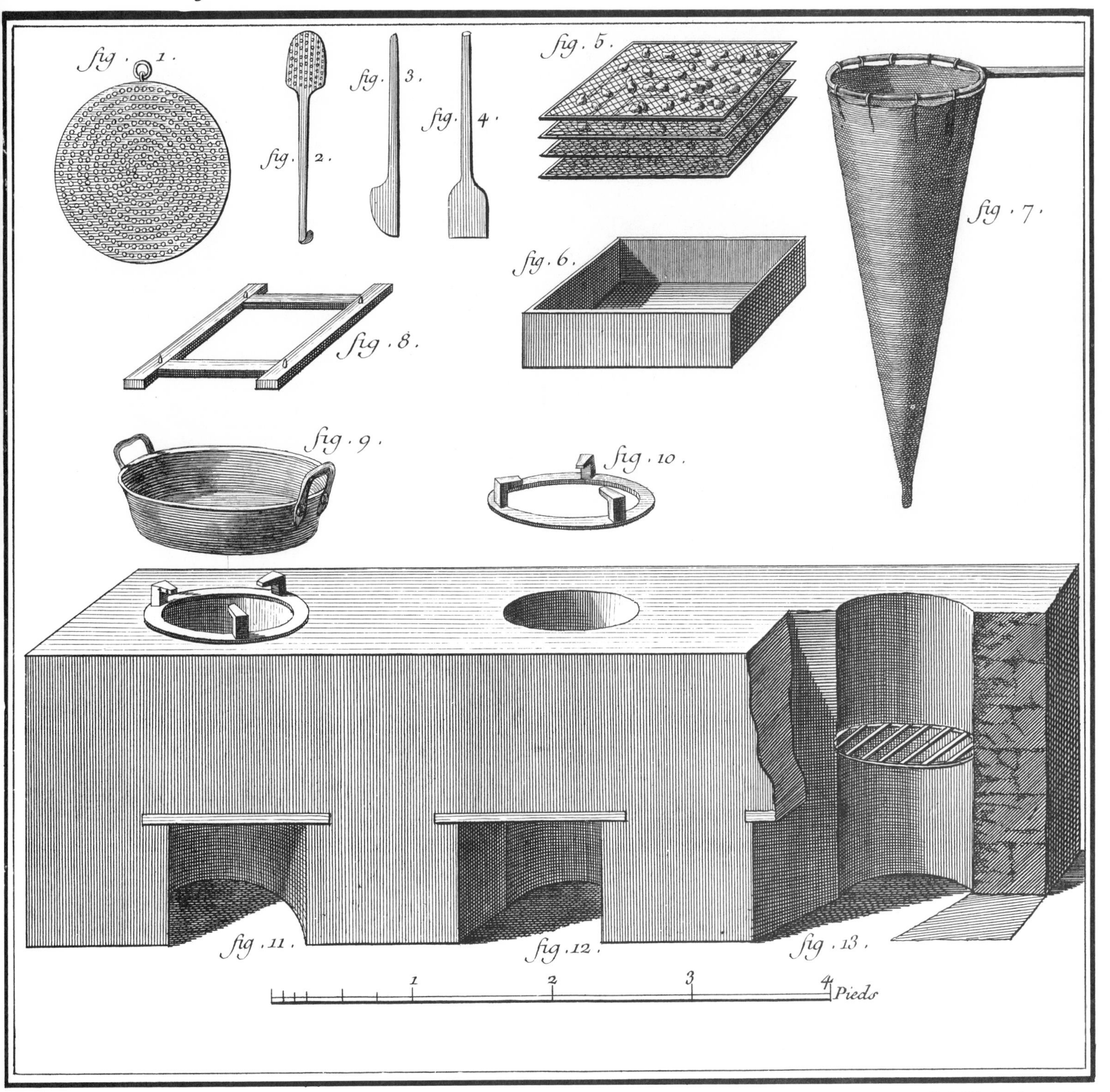
fig. 1.
fig. 2.
fig. 3.
fig. 4.
fig. 5.
fig. 6.
fig. 7.
fig. 8.
fig. 9.
fig. 10.
fig. 11.
fig. 12.
fig. 13.
1
2
3
4 Pieds

Plate 481 The Confectioner's II

A second "laboratory" produces plain (Fig. 1) and beaded (Fig. 2) sugarplums. Despite the name, they may be sugared over a core of almond, anise, pistachio, coriander, or slices of orange. In any case, the nuts are first dried and then oscillated in a mixture of gum arabic melted in a little water over the brazier. Candying consists of coating successively in sugar and gum, each layer being allowed to harden before the next is applied. The plain sugarplums are left with a final frosting of crystallized sugar. The more luxurious beaded candies (Fig. 4) are glazed in twice-boiled syrup, which drips from a funnel overhead. The worker of Fig. 3 prefers, apparently, oscillating his basin of almonds over a barrel rather than suspending it from the ceiling.

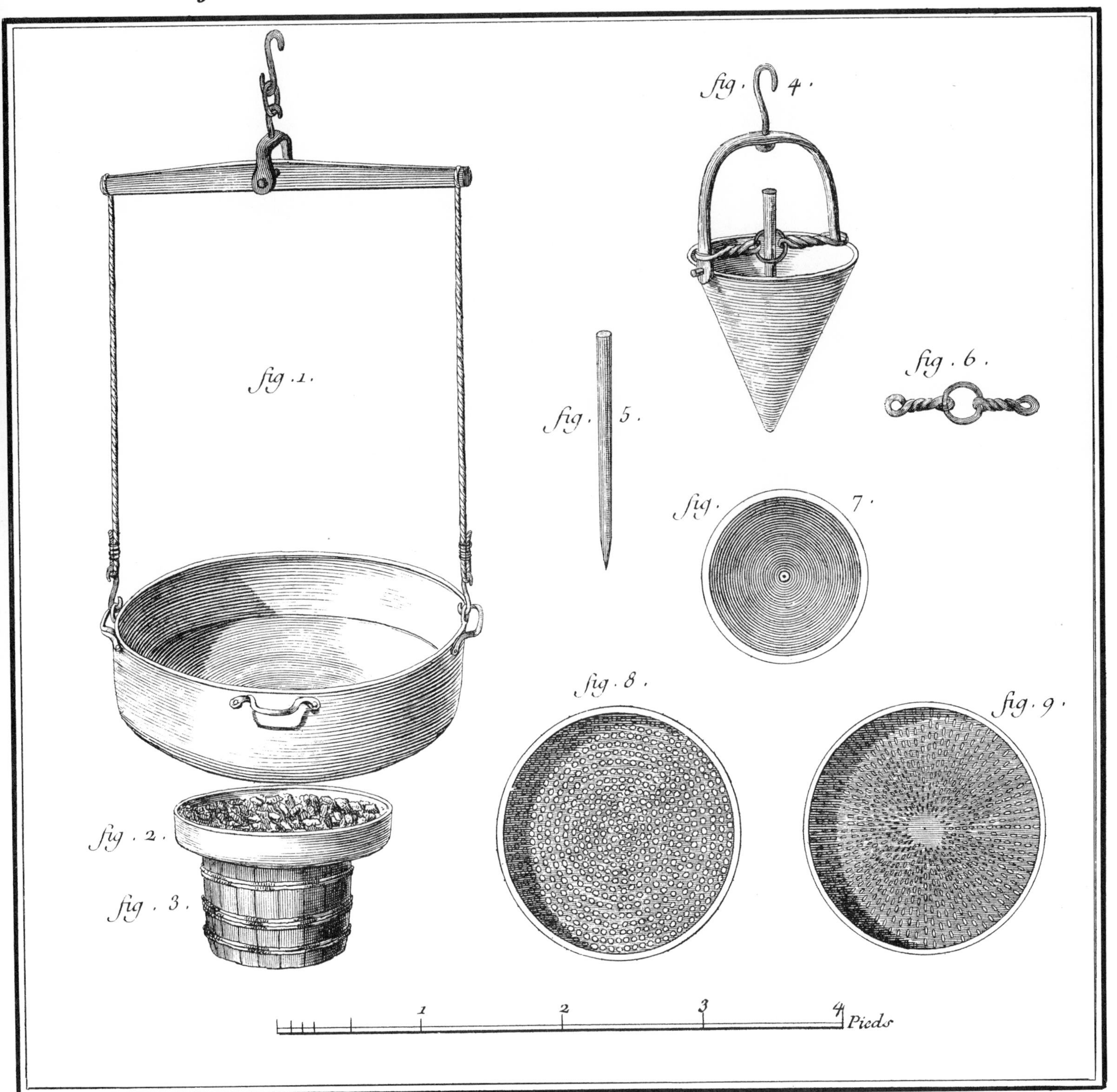
fig. 1.
fig. 2.
fig. 3.
fig. 4.
fig. 5.
fig. 6.
fig. 7.
fig. 8.
fig. 9.
1
2
3
4
Pieds

Plate 482 The Confectioner's III

This shop makes moulded sugar confections. The mortar of Fig. 1 serves for pounding gum or spice. The girl at the end of the table (Fig. 2) presses some of the designs into a rolled out candy-paste or crust. Candied flowers are the specialty of the girl seated in Fig. 3, and at the left (Fig. 4) another worker moulds the handles of a vase which is itself a piece of candy. Moulds and tools are shown opposite.

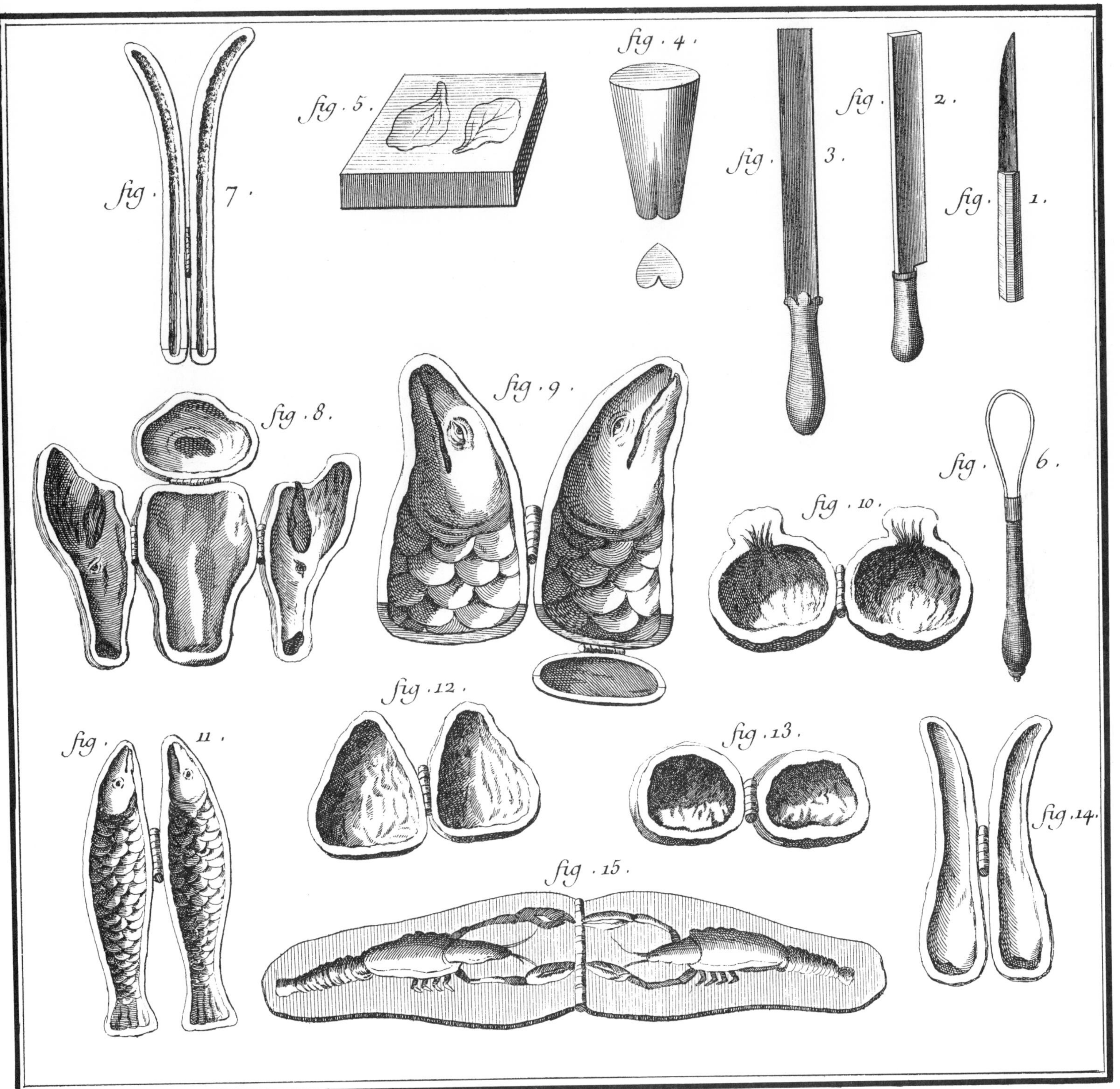

Vol. III, Confiseur, Pl. IV.

Vol. III, Confiseur, Pl. V

Plate 483 The Confectioner's IV

This shop makes chocolate drops and moulds various confections.

Vol. VIII, Patenôtrier, Pl. I.

Plate 484 Bondieuserie

Here is extreme specialization. This shop makes nothing but rosaries, the beads and ornaments of which are fashioned of bone and wood. In 18th century Paris as nowadays the dealers in such religious wares had their shops in the region around Saint Sulpice.

Plate 485 L'Envoi

If this volume accomplishes nothing else, at least it should make clear that accounts of the Encyclopedia *as an ideological enterprise conceived in praise of reason and derision of religion tell only half the story. But neither would it do to overtip the balance and to picture the Encyclopedists as some sect of solemn technicians or humorless gadgeteers. Although the plates give little scope to the sardonic spirit which informs the articles on the more sensitive subjects, it may, perhaps, be in keeping with the tone and intention of the whole to conclude, not with some magnificent new pump, not with some coke-burning blast furnace down the throat of which French industry was to be enticed (or shoved), but with this tongue-in-cheek depiction of Noah's ark; its construction and gangplank, the manner of cleaning the stalls, the loading of supplies, the airy accommodations for the animals two-by-two, the harbinger doves, one discouraged and empty-mouthed—but one flying in from the left with an olive branch to foretell, among other things, the Enlightenment.*

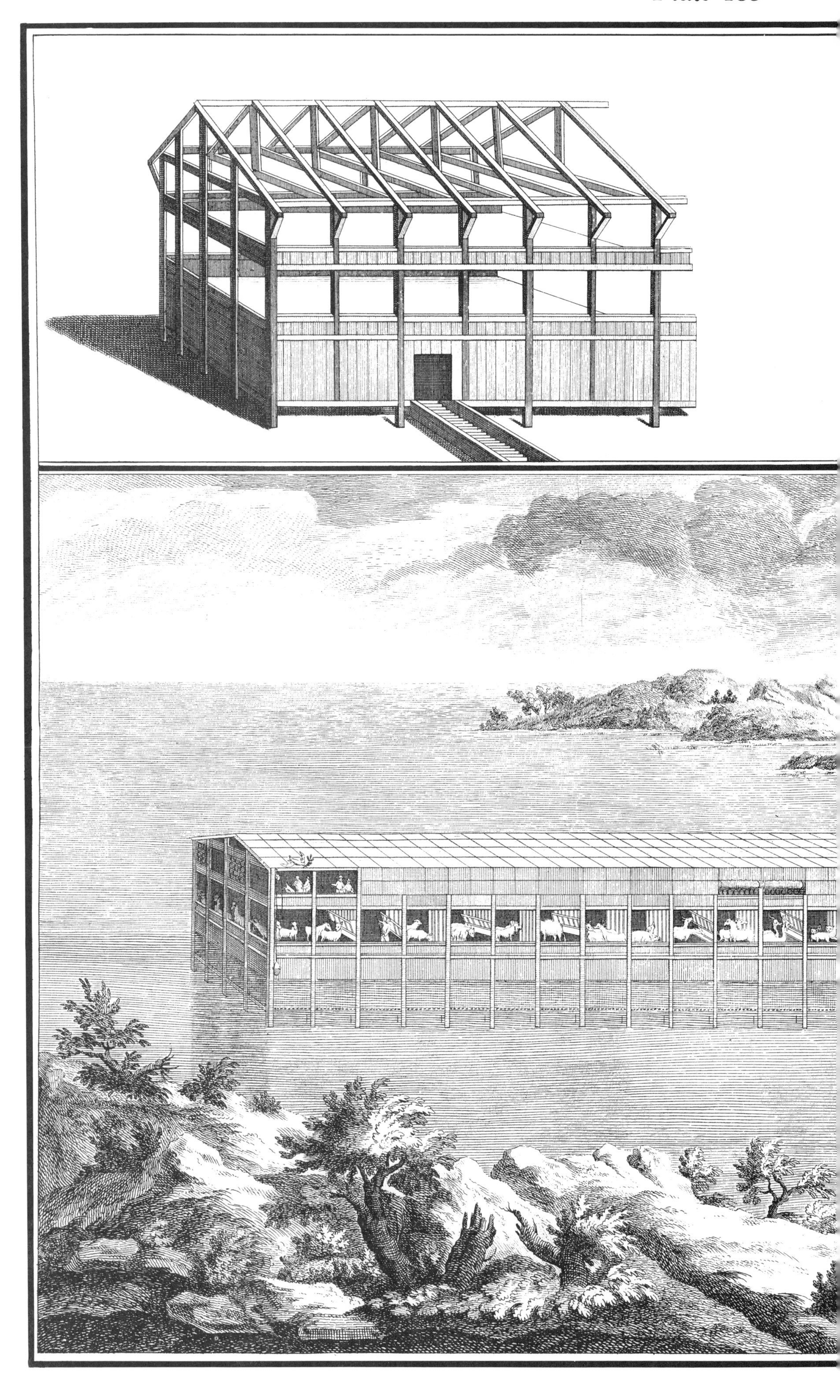

Supplement, Antiquités judaïques.

INDEX

Index

Index of Persons and Places

All roman numerals are references directed to the general introduction, pages XI-XXVI. Arabic numerals are references to the text accompanying the plate, from plate 1-485. References to (Introduction) refer to the introductory material preceeding each section.

Index

Index

Index of Subjects Illustrated in Plates

All references are to plate numbers. Major sections are capitalized.

Index of Subjects Illustrated in Plates

Index

Index

Index

Index

Index

Index

Index